The Evolution Revolution

The Evolution Revolution

Ken McNamara and John Long

Illustrations by
Danielle West

JOHN WILEY & SONS
Chichester • New York • Weinheim • Brisbane • Singapore • Toronto

Published in 1998 by John Wiley & Sons Ltd,
Baffins Lane, Chichester,
West Sussex PO19 1UD, England

National 01243 779777
International (+44) 1243 779777
e-mail (for orders and customer service enquiries):
cs-books@wiley.co.uk
Visit our Home Page on http://www.wiley.co.uk
or http://www.wiley.com

Reprinted January 1999

Other Wiley Editorial Offices

John Wiley & Sons, Inc., 605 Third Avenue,
New York, NY 10158-0012, USA

WILEY-VCH Verlag GmbH, Pappelallee 3,
D-69469 Weinheim, Germany

Jacaranda Wiley Ltd, 33 Park Road, Milton,
Queensland 4064, Australia

John Wiley & Sons (Asia) Pte Ltd, 2 Clementi Loop #02-01,
Jin Xing Distripark, Singapore 129809

John Wiley & Sons (Canada) Ltd, 22 Worcester Road,
Rexdale, Ontario M9W 1L1, Canada

Library of Congress Cataloging-in-Publication Data

McNamara, Ken.
The evolution revolution / Ken McNamara and John Long.
p. cm.
Includes bibliographical references (p.) and index.
ISBN 0-471-97406-4. — ISBN 0-471-97407-2
1. Evolutionary paleobiology. I. Long, John A., 1957–
II. Title.
QE721.2.E85M42 1998
560—dc21 98-10109
CIP

British Library Cataloguing in Publication Data

A catalogue record for this book is available from the British Library

ISBN 0-471-97406-4
ISBN 0-471-97407-2

Typeset in 11/13pt Times from authors' disks by Mayhew Typesetting, Rhayader, Powys
Printed and bound in Great Britain by Bookcraft (Bath) Ltd

This book is printed on acid-free paper responsibly manufactured from sustainable forestry, in which at least two trees are planted for each one used for paper production.

It believed in a thing called Evolution. And it said, 'All theoretical changes have ended in blood and ennui. If we change, we must change slowly, and safely, as the animals do. Nature's revolutions are the only successful ones.'

G.K. Chesterton

Contents

Preface

We stepped off the 'plane after a flight that had taken us almost as long as it takes to fly from London to Moscow. But we hadn't even left our home state of Western Australia. We had just flown from Perth in the south, to the far north of this huge state, so large that you could fit Texas into it four times. It was 1991, and being curators of palaeontology at the Western Australian Museum in Perth, we were on our way to the Canning Basin to hunt for fossils in what are arguably the most spectacular fossil reefs of their age anywhere in the world. We had been drawn to these 370 million-year-old limestones for two reasons. One was the chance for John to search for some of the most exquisitely preserved fish you could ever hope to find; the other was for Ken to fulfil his passion of looking for trilobites. These extinct, distant relatives of insects and crayfish had been thought to be very rare in this part of the world – that is until amateur fossil sleuths Gutha Rokylle and her partner Esko had made some startling discoveries.

Wandering some of these limestone ridges, once under ancient seas, but now surrounded by seas of prickly spinifex grass, Gutha and Esko had started to become fascinated by the fossils that dotted the rocks. After sending me a few specimens for identification I suggested that they look for these strange little fossils called trilobites. Off they went and within days had discovered an immensely rich, previously unknown fauna that straddled one of the most significant periods in Earth history. Known as the Frasian–Famennian boundary, it is one of the five big mass extinction events that punctuated the evolution of life on this planet during the last half a billion years. What these trilobites had the potential to reveal was how the faunas had been affected by this big event. How had the different lineages of trilobites evolved? Could we glean from this any idea of whether these were environmental stresses leading up to the boundary? Or were the trilobites and the rest of the fauna just happily living, evolving and dying, but were then suddenly wiped out, perhaps by a large meteorite, as some have argued.

Herein lay two good reasons for us to travel so far away from our cosy homes in Perth to the desert wilds of the Kimberley region of W.A. – not only the excitement of collecting these fossils and, in all likelihood, the possibility of finding forms new to science, but also finding the evidence to answer many of these questions. After nearly one weeks' collecting we had found thousands of specimens; enough, surely, to solve at least some of these scientific enigmas. John, however, was champing at the bit. Just 5 kilometres away up the valley were sediments deposited in deeper water off the front of the ancient reef. Entombed within limestone nodules lay perfectly three-dimensionally preserved fishes, still with their original bone. Here the excitement of the discovery was delayed. For it was not until the rocks were immersed in weak acid back in Perth and the rock slowly dissolved away, that the identity of the entrapped fishes could be revealed. Thus, almost every day at work there was the potential for a new species of fish to be discovered in the laboratory. It made coming to work on Monday mornings a real treat to see what the acid baths had revealed over the weekend.

So, herein lies the essence of what this book is all about – fossil discoveries, made both by professional palaeontologists as well as by dedicated amateurs, and how these new discoveries can shed new light on the patterns and processes of evolution that have been ticking away on this planet for more than three and a half billion years.

Evolution is, without doubt, a very complicated process. We scientists argue amongst ourselves about how it actually occurs. And so we should. For it is the only way in which science can progress. Yet, despite an overwhelmingly vast scientific literature documenting the patterns and processes of evolution, the extent to which much of this finds its way into the public domain is rather limited. This is why we have written this book, to disseminate much of the latest information that has come to us through academic channels. We have mulled over it, added a dash of our own spicy ideas and peppered viewpoints, and now present it in what is hopefully an easily digestible and tasty format for the enthusiastic reader.

The information that we have used comes from the three-and-a-half-billion-year-old long-playing fossil record. The fossil record reveals patterns in the evolution of life over this vast tract of time. The application of modern theories of evolution enables the fossil record to make its own, unique contribution to our understanding of some of the mechanisms of evolution. Throughout this book we hope to share with you some of the excitement and fascination that comes not only from making fossil discoveries, but also of finding out how they contribute to our understanding of biological evolution. Palaeontology, the study of fossils, has certainly come of age in the last two decades, and has taken its place within those scientific disciplines involved in the formulation of ideas on how evolution works. The dramatic unravelling of the fossil record by continual new discoveries has played a major role in stimulating new ideas on the rates of evolution, the impact that mass extinctions have had on the pattern of evolution, and on our understanding of the underlying processes and mechanisms of evolution.

Even as we write this, we know that many of our colleagues around the world are currently scouring places like dry, dusty deserts, towering mountain ranges and precipitous sea cliffs and making exciting new fossil finds that will add a few more pieces to the great jigsaw of life. The result of such collecting has been a startling increase in the rate at which new and important finds are being made. Compared with Darwin's day, we now have millions more fossil specimens in world collections, neatly filling in many of the major gaps in the evolutionary tree of life. Gaps that troubled Darwin when he attempted to incorporate the fossil record into his concept of evolution by natural selection. In our view we are right now experiencing a revolution in the way in which the fossil record is contributing to our understanding of the patterns and processes that have crafted the living world as we see it today.

Our book is a series of 18 essays, each dealing with some aspect of the fossil record and how it has contributed to our understanding of evolution. Because we have written this book from a very personal perspective, often sharing our own experiences, we have written parts in the first person singular. To satisfy your curiosity, you might like to know who has contributed to each chapter, although both of us have had some input into all the chapters. The chapters on early life, the earliest animals, the Cambrian Explosion (Chs 1–3), colonisation of land (Ch. 5), evolution of insects (Ch. 7), evolution in plants (Ch. 9), early bird evolution, big dinosaurs (Chs 13, 14), impact of predation on evolution, human evolution and classic examples of evolution revisited (Chs 16–18) were written by Ken McNamara. Chapters dealing with the earliest vertebrates (Ch. 4), early fish evolution (Ch. 6), the first tetrapods (Ch. 8), marine and flying reptiles, and polar dinosaurs (Chs 10–12) and mammals (Ch. 15) were written by John Long. Chapter 14, dealing with how dinosaurs grew to be so big, is based on research that we have jointly been carrying out in recent years.

We like to imagine Darwin's spirit smiling down on palaeontologists for revitalising his favourite subject. Despite this book being only a small contribution towards dispelling the misunderstanding that still surrounds the topic of evolution, we hope that it will excite and stimulate readers of all ages and backgrounds to promote and discuss these evolutionary ideas within their communities, schools and universities. *Viva la Revolution!*

Ken McNamara and John Long
Perth, 1997

Acknowledgements

Books like this cannot be written in isolation. Insomuch as we have drawn heavily on the work of our many colleagues around the world, we have also drawn on their various areas of expertise to assist us with many of the chapters. One of the satisfactions with writing this book has been the unstinting help given to us by all of those colleagues we have approached. Despite the fact that we are mere interlopers into fields where we have had little direct research experience, all of our colleagues, without exception, have given freely of their time and experience, promptly and encouragingly. Others have furnished us with up-to-date references that we had missed or were unable to find, while other colleagues and friends have read a few or all of the chapters or helped us with the challenging task of coming up with chapter titles. Naturally, any mistakes or misinterpretations are our responsibility. To all these people we offer our sincere thanks: Per Ahlberg, Mike Archer, Alex Baynes, Alex Bevan, Jenny Bevan, Luis Chiappe, Mike Coates, Simon Conway Morris, Rob Craig, Arthur Cruickshank, Greg Edgecombe, Tim Flannery, Jim Gehling, Philippe Janvire, Mike Lee, Steve McLoughlin, Ralph Molnar, Tom Rich, Alex Ritchie, Andrew Rozefelds, Barbara Rye, Leo Salgado, Pat Vickers-Rich, Gavin Young and Yu Wen. We also wish to thank Isabelle Strafford, Louise Portsmouth and Karin Fancett at Wiley for their help in carrying our manuscript through to publication.

But most of all we wish to express publicly our heartfelt thanks to Danielle West for her many beautiful drawings that have done so much to improve the appeal of this book.

1

We are the Champions!

The long-playing microbial fossil record

It was a billion years since the Earth was born. Volcanoes belched deep into the air. But this was not an air you could breathe, for it was an air devoid of oxygen, consisting mostly of nitrogen, sulphur dioxide and carbon dioxide. The land was a bare eruption of black, rubbly wind-swept rock. Neither plant nor animal clung to its surface. Winds lashed the seas into towering castles of foamy waters. But nestling in one quiet spot, on the edge of a newly formed continent, in a shallow embayment, something was stirring beneath the water: something too small to be seen with the naked eye. In shallow marine embayments the planet's first construction workers were busy. For here, in the relatively quiet waters, some of Earth's earliest life forms, tiny microbes, flourished in what we would call a harsh environment. Yet, to these early bags of DNA, it was paradise. Trapping and binding the sediment, these primitive life forms were constructing edifices of gargantuan proportions: domes of cemented sediment, millions of times larger than their individual cell sizes; monuments to life's first attempts to manipulate the environment.

Even today this site, tucked away deep in the Pilbara region of Western Australia, is still considered by us to be a very harsh environment. Wander too far away from your vehicle without water in the height of summer and you can be dead in a couple of days. To the microbes, it no longer matters – they are long dead. Yet, amazingly, their fossilised remains have been found entombed in these sediments, formed so long ago, along with the sedimentary cities that they created. Fossilisation is not just within the purview of bones and shells.

We are able to piece together this scene of early life on Earth thanks to the work of a number of geologists and palaeontologists. Pioneers in reconstructing this ancient world in the Pilbara were John Dunlop and Roger Buick from the University of Western Australia. They were carrying out mapping in the area in the late 1970s at a locality in the Pilbara region of northwestern Australia called North Pole[1]. Australians have a strange sense of humour, for the closest town, Marble Bar, boasts having gone for 162 consecutive days with daytime maxima in excess of 100°F. The rocks that Dunlop and Buick were mapping are amongst the oldest known on Earth. More recently Buick and other geologists from the University of Western Australia[2] have identified what they consider to be emergent continental crust that formed in this area about 3500 million years ago. In other words it is possible to stand on the actual land surface that existed at that time. It is now represented by an unconformity with volcanic and sedimentary rocks that made up the ancient land surface, covered by younger sands, cherts and basalts. Even a possible fossil soil can be recognised.

When Dunlop and Buick first looked at one of the domed mounds of stratified rock, rich in chert and barite, they made the mental quantum leap of equating the structures with similar domed features known as stromatolites, that are still forming today just a few hundred kilometres away in Shark Bay, on the Western Australian coast. But the domes they were looking at on a hot, rugged, spinifex-covered hillside had seen 3500 million summers pass since their microbes were last active.

Stromatolites (also sometimes known as microbialites) are rare today. Where they do occur they are found in a wide range of aquatic environments, usually ones hostile to anything except stromatolite-building microbial communities. In the sea they occur from the subtidal to supratidal, and in hypersaline to normal salinity seawater. Non-marine stromatolites have been found in streams, lakes, thermal springs and even beneath frozen lakes. Stromatolites are built by complex communities of microorganisms, often involving many tens of species, and with population densities of more than 3000 million individuals per square metre. Living stromatolites are mainly constructed by photosynthetic cyanobacteria (formerly known as 'blue–green algae') and may grow up to tens of metres in height. Other photosynthetic microbes also contribute to their growth, including diatoms, green algae, red algae and other kinds of bacteria. Stromatolites usually grow in areas where the water has elevated salinity or alkalinity, low nutrient levels, periodic dessication and other seasonal changes, elevated or decreased temperatures, precipitation of mineral material during growth, and strong wave or current action. Because they are usually constructed from calcium carbonate, stromatolites are generally found in areas where the water is rich in calcium and bicarbonate[3].

Microbial communities construct stromatolites in two main ways. One way is by trapping sediment with a sticky film of mucus that each cyanobacterial cell secretes, then binding the sediment together with mucus and growth of the cyanobacteria over the grains. Finally calcium carbonate is precipitated from the water and cements the grains. Being photosynthetic and motile the cyanobacteria move toward the light, so they can keep up with accumulating sediment. This means that they always remain on the outer surface of the stromatolite. A second way that stromatolites are constructed is by the cyanobacteria precipitating their own carbonate framework, with little incorporation of sediment. This produces a stromatolite with little or no internal layering, and a characteristic 'clotted' internal structure. This type are often called 'thrombolites'.

The twentieth century discovery of stromatolites nearly 3500 million years old was the culmination of over 20 years of work by palaeontologists who had searched the ancient Precambrian rocks that had been thought to be lacking of any evidence of former life. In the 1930s the English palaeobotanist A.C. Seward, writing in his book *Plant Life through the Ages*, had been convinced that suggestions for the existence of Precambrian fossils were folly, making 'a demand upon the imagination inconsistent with Wordsworth's definition of that quality as "Reason in its most exalted mood" . . .'. The great American evolutionary biologist, G.G. Simpson, was of a like mind. He couldn't conceive that microscopic microbes of any shape or description would survive the fossilisation process, thus making any search for them in Precambrian rocks futile. One palaeontologist of the early twentieth century, though, who did consider that life had existed in Precambrian times was Charles D. Walcott. Famous for having discovered the fossil riches of the Burgess Shale (see Chapter 3), Walcott actively described stromatolites and in 1915 described microscopic bodies that he thought might possibly represent the fossilised remains of bacteria[4].

J. William Schopf of the University of California at Los Angeles, one of those in the forefront for the search for evidence of Precambrian

Cauliflowers from the dawn of time? No, these are stromatolites, which represent one of the oldest life forms on this planet, formed by microbial communities dominated by cyanobacteria, dating back some 3.5 billion years. Today they are still thriving in places like Shark Bay in Western Australia.

life, has noted that since the initial reporting in three papers published in the journal *Science* in 1965, more than 3000 occurrences of microfossils have been found in almost 400 different Precambrian geological formations[5]. Moreover, many other sites have been found around the world where stromatolites are still forming, in a wide variety of environments, being constructed by a wide range of microbial communities.

Each one of these precious sites is a natural prehistoric laboratory that gives us a privileged glimpse back in time, revealing what conditions might have been like when the Earth was still a child. Studies undertaken during the last few years on both actively forming and fossil stromatolites and their associated microbial communities is revealing that for nearly three-quarters of its existence the Earth was a complex world dominated by organisms so small that they make even the smallest of the animals that evolved late in Earth history (see Chapter 2) – the new kids on the block – look like monsters of gigantic proportions.

The oldest fossil microbes associated with the stromatolites were first recovered from cherts at North Pole[6]. What was stunning about these first finds of primitive microbes was that they bore a striking resemblance to some bacteria still living today. Further work, principally by Schopf, has uncovered a surprising diversity of forms, suggesting that the evolution of such microbes had probably been going on well before 3500 million years ago.

Finding fossilised microbes in such archaic rocks is no mean feat. Most rocks more than 3000 million years old have been altered beyond recognition, destroying any fossils that might have been entombed within them. But the Pilbara region of Western Australia has been geologically stable for an immense period of time, to such an extent that original sediments are still preserved, horizontally layered, essentially unaltered. Schopf's studies to date have revealed 11 different types of microbes[7]. Most were recovered from the 3450 million-year-old Apex Chert, located about 12 km west of Marble Bar. Perhaps the most amazing aspect of research on these early microbes is the realisation of just how similar they are to living cyanobacteria. If these bacterial filaments really were cyanobacteria, then oxygen-producing photosynthesisers had evolved extremely early in Earth history. Although rare, microfossils older than 2.5 billion years have been discovered in just a few other sites: the 3465 million-year-old Towers Formation and the 2750 million-year-old Tumbiana Formation, both in Western Australia; and the 3450 million-year-old Swaziland Supergroup in South Africa.

Surprisingly, these ancient filaments are not only very similar morphologically to their living counterparts, showing a remarkable diversity so early in their evolutionary history, but as such microbes go they are actually quite large. Intuitively we would expect so ancient and primitive fossils, simple prokaryotic cells consisting of little more than a bag of DNA, to be extremely small. However, Schopf has shown that taxonomically the filaments from the Apex Chert are more diverse (principally in size) than 92% of the 126 other assemblages of similar microfossils found throughout the Precambrian. Indeed, the assemblage is more than twice as diverse as the average assemblage. As Stephen J. Gould of Harvard University has argued[8], we should not be lulled into the assumption that evolution necessarily proceeds from the simple and small to the large and more complex. Evolution is about increasing diversity, both up and down. Many other filamentous microfossils found in later Precambrian rocks are much smaller and more simple than these earliest known examples.

Such palaeontological discoveries push back the origin of life to its very limits. If such a diverse microbial assemblage was in existence about 3.5 billion years ago, then how far back can we go in our quest for life's origins? Whether older microfossils will ever be found is debatable. The limiting factor will be the discovery of older, unaltered sediments. Yet, clues to early forms of life are coming from other sources, providing indirect evidence of the existence of life on the planet early in its history.

Recent research on the carbon-isotope composition of carbonaceous inclusions within grains of the mineral apatite (a calcium phosphate) from Isua in Greenland have suggested that life existed at least 3800 million years ago[9]. Isotopically light carbon indicates biological activity although it has also been argued that metamorphism of carbonaceous matter in sedimentary rocks can also produce this effect. Such isotopically light carbonaceous inclusions have been found in apatites formed in banded iron formations deposited about 3250 million years ago in Western Australia, and in some more than 3700 million years old in West Greenland. Similar isotopically light carbonaceous inclusions have also been found within the 3450 million-year-old rocks from near Marble Bar. If the Earth was subjected to heavy bombardment by meteorites for about 600 million years, from its birth until about 4000 million years ago, to such an extent that it inhibited the evolution of life, then we are down to a very narrow window of time during which life could have evolved if we accept that life was present 3800 million years ago.

The indication that very early bacteria were photosynthetic has been used as argument for a biogenic origin to the vast deposits of iron ore that are scattered around the world in very ancient Archaean rocks. Although the banded iron formations in Greenland may be more than 3700 million years old, in other parts of the world most are between 2700 and 2000 million years old. At this time the atmosphere was devoid of oxygen, being mainly composed of nitrogen and carbon dioxide. To organisms that had evolved under these anoxic conditions, oxygen was a poisonous waste product to be disposed of. To be able to do this they possessed oxygen-mediating enzymes. At this time in Earth history the seas contained large quantities of dissolved iron. Combination of the oxygen waste product with ferrous iron dissolved in the water resulted in the production of insoluble ferric oxide that was precipitated as banded iron formations – the seas simply rusted.

Once all the dissolved ferrous ions in the seawater had been converted into ferric oxide, the oxygen that continued to be produced by the photosynthetic activity of the cyanobacteria accumulated firstly in the seawater, and then in the atmosphere. This would probably have been disastrous for all those anaerobic microbes that had evolved in a world free from oxygen. The only microbes able to survive would have been those that could find niches where oxygen didn't exist, or those that could metabolise it in some way. Some evolved special enzymes that enabled them to exist in an oxygen-rich environment. Whether or not the very early banded iron formations in Greenland were similarly formed by the action of photosynthesising bacteria has been the subject of much debate. However, the recent discovery of indirect indications of life at this time seem to indicate that even these earliest microbes might have been capable of photosynthesis.

Other suggestions for the mechanisms that formed banded iron formations include photodissociation of water in the atmosphere, or photochemical reactions that extract oxygen from seawater. Experiments with living bacteria have indicated that some forms of anoxygenic phototrophic bacteria are also able to oxidise ferrous to ferric iron. This strengthens the argument for the role of early bacteria in the production of BIFs. In all likelihood, different processes were involved in their formation. However, the role of early microbes was probably crucial. Without the activity of photosynthetic microbes, life would not have progressed in the way that it has, and we, and our cabbages and daffodils, would not exist.

What the first life was and how it formed is obviously the 4600 million dollar question. Although life today is DNA based, it has been argued that a pre-DNA world existed, when life forms were based on the structurally more simple RNA (ribonucleic acid). Because the role of RNA is to translate information from the DNA molecule, in order to furnish the blueprint for the formation of proteins, it is a perfect vehicle for storing and transmitting information. RNA also has the ability to act as a catalyst, enabling new RNA to be formed from its constituent nucleotide building blocks. Presumably some sort of membrane would have been needed to hold the RNA together so that it could function like a very basic cell.

There have been countless suggestions made as to how and where the basic nucleotides first came together to form something as complex as an RNA molecule[10]. Jack Szostak of the Massachusetts General Hospital has suggested that amongst the vast numbers of nucleotides dissolved in water in small pools, a few would have combined and developed the ability to replicate themselves. With colleagues he tested this idea by taking 100 to 1000 trillion RNA molecules, each only about 200 nucleotides long, and selected out the few that made copies of themselves – simulating natural selection at work. Using protein enzymes they made millions of copies and ended up producing new RNA cutting ribozymes. These are the catalytic RNAs that had the ability to cut out portions of their own nucleotide sequence, as well as stick pieces together.

Where did the building blocks of RNA first come from? To understand this we need to go back even further in time, to when there were just free nucleic acids: adenine, cytosine, guanine and uracil. But where did they come from? Stanley Miller's famous experiments in the 1950s, in which an electrical spark passed through ammonia, hydrogen, water vapour and methane (all likely components of Earth's early atmosphere) and produced amino acids, showed one means by which life could have been kick-started. In addition to the amino acids, large quantities of adenine and guanine were produced. The problem was that, until recently, cytosine and uracil had not been able to be synthesised in a similar fashion. However, in 1995, Miller, along with Michael Robertson at the University of California at San Diego, realised that these two nucleic acids could be produced in a primordial soup by having high quantities of urea in the soup. Reacting with cyanoacetaldehyde, which is produced by the spark and hot air, large quantities of cytosine and uracil were formed. Miller's view is that in small pools, where evaporation levels were high, urea would be concentrated. This is their 'drying lagoon hypothesis'.

Having amino acids is one thing. Creating great chains of RNA to form life is quite another. How and where could this occur? An idea favoured in recent times is that clays mediated this reaction. Jim Ferris of the Rensselaer Polytechnic Institute in Troy, New York, found that by adding the positively charged clay montmorillonite to a solution containing negatively charged adenine nucleotides, RNA 10–15 nucleotides long could be formed. By adding more nucleotides, RNA molecules 55 nucleotides long grew.

Opponents of the idea of an RNA world have asked where did the necessary energy to make RNA molecules come from. John Glover of the University of Western Australia has proposed that the energy source may have been radioactivity. He and Berger Rasmussen noticed that grains of radioactive monazite (phosphates of cerium, lanthanum, thorium and neodymium) in sandstones are often enveloped by a solid, highly polymerised bitumen-like organic compound. They have suggested a model whereby monazite grains were concentrated along the shores of shallow pools, where evaporation levels were high. This would have enhanced the effect of the radioactivity, which would have provided the source of energy needed for life to first emerge from the primordial soup[11].

But are clays the only inorganic mediating agents that could have been the catalyst for the evolution of life? Günter Wächtershäuser of the University of Regensberg in Germany thinks not. His more radical idea is that these early organic compounds were synthesised and polymerised from inorganic constituents, on positively charged minerals like the iron sulphide, pyrite. Negatively charged organic compounds were attracted to the pyrite. As energy was generated during the formation of iron sulphide, more organic molecules were constructed[12].

In recent years another view of where life might have started on Earth has begun to emerge. Rather than forming in some pool of organic molecules, it has been suggested that life began in environments that we would consider to be very unpleasant – boiling, sulphur-laden pools, or mineral-rich, deep-sea volcanic vents. In such environments it has been argued that life could have started as far back as 4.3 billion years ago, or even earlier. By analysing ribosomal RNAs, nucleic acid sequences that occur in ribosomes, microbiologist Carl Woese has argued strongly that many of these first 'thermophiles' belong in a group that he calls 'Archaea' (formerly known as 'Archaeobacteria'), and that these, along with the Bacteria and Eukarya (multicellular animals and plants) comprise all life on Earth. The argument goes on. Some say that even these heat-loving Archaea are relatively complex, and that they themselves may have been derived from even simpler life forms that developed in cooler waters. It is, of course, possible that they may well not have been the only organisms on this youthful Earth. But they may have been the only ones capable of withstanding the massive impacts that occurred during the bombardment episode.

But did the building blocks of life really start on Earth? An idea gaining more support, particularly with the 'discovery' of what are purported to be fossils of filamentous bacteria in a Martian meteorite recovered from Antarctica, is that the building blocks for life may have originated in outer space. Some meteorites, such as the Murchison meteorite, more than 100 kg of which fell in Victoria, Australia, on 28 September 1969, contain abundant amino acids – the building blocks of life. Amino acids have amino (NH_2) and carboxyl (COOH) groups. Eliminate the water molecule and a peptide bond is formed. Polymerisation of amino acids by the formation of a series of these bonds can produce proteins.

Perhaps what is so surprising about this vast period of time which we call the Precambrian, that occupied almost the first nine-tenths of Earth history, is the overwhelming dominance of bacteria and their handiwork – the stromatolitic structures that some of them created. As we discuss in the next chapter, more complex cells, the first algae, probably evolved by at least 2000 million years ago, and the first animals may be as much as 1000 million years old. However, they played a very small part in the evolution of Earth's early ecosystems, compared with the last 600 million years. Research into fossilised Precambrian microbial filaments and stromatolites is revealing the patterns of evolution and changing diversity of these forms through much of the Precambrian. The fossil record reveals not only their rise to dominance during much of the Precambrian, but also their decline and fall as more complex, multicellular life evolved rapidly in the later part of the Precambrian, as we discuss below.

From the first simple domes, a range of stromatolite morphologies had evolved by the end of the Archaean, 2500 million years ago. These include wavy laminated, nodular, domical, pseudocolumnar, columnar, conical and spherical (oncolitic) forms. Columnar stromatolites were rare, but some had started to produce more complex, branching structures. Most of these varieties of stromatolites have persisted to the present day and can still be seen in quiet inlets and coastal lakes in some parts of the world – surely the ultimate 'living' fossils. Like stromatolites today, it is likely that filamentous cyanobacteria were the dominant stromatolite builders in the Archaean. However, the increase in complexity of the structures that these early microbial communities built probably reflects an increase in the complexity of the microbial communities themselves. We know that today, in addition to cyanobacteria, algae can also play an important role in stromatolite construction in certain environments. So it probably was with some of the more complex Late Archaean and Early Proterozoic stromatolites[3].

Archaean stromatolites are generally pretty rare. This may be a reflection more of geological factors, rather than biological ones. During most of the Archaean Eon (3800 to 2500 million years ago), sedimentation in marine environments was characterised by rapid periods of deposition. Under such conditions, phototrophic, motile microbes would have had difficulty in maintaining their position at the sediment/water interface, making stromatolite growth very difficult. The stable shallow water environments that are more conducive to stromatolite growth, as well as preservation, were less common at this time. Another factor may have been the relatively uncommon occurrence of carbonates. However, the development of larger, more stable, shallow water environments near the end of the Archaean was ideal for the proliferation of stromatolites during the following Proterozoic Eon (2500 to 570 million years ago). In such environments carbonate sedimentation is more likely to have occurred, and without its availability stromatolites rarely occur.

It was during the Proterozoic Eon that stromatolites reached the pinnacle of their development, both in abundance and diversity. As well as the simple forms of stromatolites established in the Archaean, conical, columnar and branching columns became common. Many have been described, given scientific names and found to have restricted time-ranges. This high diversity has stimulated research into the adaptations of different stromatolite morphotypes to different environments. Kath Grey and Alan Thorne of the Geological Survey of Western Australia have shown that the restriction of different types of stromatolites to certain sediment types reflects deposition under different environmental conditions. This allows reconstructions of very ancient palaeoenvironments. One example is in the 2000 million-year-old Duck Creek Dolomite in the great treasure-trove of Precambrian stromatolites, the Pilbara region of Western Australia. Here Grey and Thorne recognised two types of stromatolites: a columnar variety called *Pilbaria*, which they think grew in shallow lagoons and into the lower intertidal zone; and another, broader-domed, digitate variety known as *Asperia*. This form appears to have inhabited pools of water in high intertidal or supratidal regions. These two types of stromatolites can be recognised in many Proterozoic rocks. We can reconstruct these ancient environments by turning to some of the living analogues[13]. Living stromatolites that are remarkably similar to these 2000 million-year-old ones can still be seen in some Western Australian lakes today.

Over the last 20 years, detailed documentation of the Precambrian stromatolite record has revealed that during the Proterozoic there were two major periods of diversification, one 2500 to 1650 million years ago, the other 1350 to 1000 million years ago. From the first period 176 different forms of stromatolites have been described, and from the second, 342. Following the last diversity high, stromatolite diversity plummeted to the end of the Precambrian[3].

The first period of diversification in the Early Proterozoic is arguably the most important adaptive radiation that affected both the stromatolites and the microbes that made them. This was a time when oxygen levels in the atmosphere began to increase. Cyanobacteria probably diversified rapidly and were widely dispersed. The evolution of new types of stromatolites was probably an outcome of the evolution of new types of cyanobacteria that could move faster and grow quicker. This would have influenced how they interacted with the sediment. Why stromatolites underwent a second burst of diversification in the Late Proterozoic is less easily explained. Maybe the steadily rising oxygen levels in the atmosphere played a part. More likely it was the fact that the eukaryotes (firstly in the form of algae) became more common.

Most biologists now accept that life can be divided into three basic groups (known to some as 'empires'). One contains the prokaryotic Bacteria (including cyanobacteria), another the prokaryotic Archaea (including halophiles), and the third the Eukarya, eukaryotes, which include all plants, animals and fungi. Eukaryotic cells are more complex than the prokaryotes, containing a nucleus and other organelles, like mitochondria. But Archaea and Bacteria are enormously more prolific. While only about 5000 species of bacteria have been described, researchers consider that millions of species exist today. In just one gram of soil that he investigated, Norwegian scientist Vidgis Torsvik estimated that there were at least 10 000 different types of bacteria!

Competition between the eukaryotic algae that lived in and on the sediment, and the long-established cyanobacteria may have promoted an increased stromatolite diversity during the Precambrian in a manner that we do not fully understand. Research on the living stromatolites in Shark Bay in Western Australia by Philip Playford and Tony Cockbain of the Geological Survey of Western Australia has shown that, depending on their depth of formation, some stromatolites have significant algal components.

The sharp decline in stromatolite diversity during the Late Proterozoic and Early Palaeozoic has long been attributed to the impact of grazing and burrowing by the animals that were evolving at this time. One argument has been that competition from these animals, in particular grazing pressure, was the critical factor in this decline in diversity. Actively growing stromatolites, it has been argued, flourish in Hamelin Pool in Shark Bay, a very hypersaline water, because few grazers can cope with this highly saline seawater. Thus the absence of animals in the early part of the Precambrian explains the dominance of stromatolite-building microbial communities. The Late Precambrian decline is said to be a consequence of the evolution of animals that grazed on the stromatolites. The problem has been, however, that fossil evidence for animals older than about 600 million years is lacking (see Chapter 2) . Recently, however, Greg Wray of the State University of New York at Stony Brook and colleagues, have suggested on the basis of molecular evidence that animals existed as much as 1200 million years ago[14]. If this was the case, there is a reasonably close correspondence between this suggested time of their appearance and the beginning of the decline of the stromatolites on quite a spectacular scale. From a dizzy peak between 1000 and 1300 million years ago, stromatolite diversity dropped to about 75% of this level between 1000 and 700 million years ago, falling more rapidly to just 20% at the end of the Precambrian. However, another study recently undertaken[15] suggests that the molecular evidence for the divergence times of the basic animal groups is close to that indicated by the fossil record, at about 670 million years ago. This is clearly an area of investigation that is currently causing a great deal of interest.

But was the adverse effect that animals were having on these cyanobacterial-dominated microbial communities purely from grazing pressure? The brackish water in which the stromatolites of Lake Clifton in Western Australia are struggling to grow also provides a test for the long-held view of the importance of grazing in the decline of Precambrian stromatolites. At the northern end of this 21.5 km long lake is a reef about 30 metres wide made from coalesced stromatolites, each up to 1 metre across. This reef extends for more than 8 km. Unlike at Hamelin Pool, the Lake Clifton stromatolites are populated by a diverse invertebrate fauna containing many grazers, mainly crustaceans (isopods and amphipods) and molluscs (gastropods and bivalves). The stromatolites are a source of both food and refuge for these animals. Yet the structures have still actively grown.

However, a far more insidious threat hangs over these particular stromatolites – not from grazing, but from human-induced increases in nutrient levels in the groundwater that feeds the lake.

The heavy use of superphosphate on agricultural land around the lake over the last few decades has resulted in a decline in stromatolite growth due to pronounced increase in phosphate levels in the groundwater that feeds the lake. This has resulted in excessive growth of the alga *Cladophora* on the submerged stromatolites. The long-term consequence of these changes to the groundwater is likely to be the death of the microbial community and the end of stromatolite growth. A similar effect seems to be occurring in Walker Lake, Nevada, USA, where stromatolites are covered by *Cladophora* during the summer months, again thought to be caused by increases in nutrient levels from agriculture and grazing. Could it therefore be possible that a major reason for the decline in stromatolite diversity in the Late Precambrian, apart from increased levels of grazing, has been changes in nutrient levels in the sea? Living stromatolites are known to flourish in environments that are low in nutrients. Increase the nutrient levels and out go the cyanobacteria and the microbial communities that construct the stromatolites, as algae outcompete the cyanobacteria.

If I may beg your indulgence, I will end this chapter by blowing the trumpet for the part of the world in which John and I both live and work: Western Australia. Not only were the world's oldest stromatolites and the microbes that constructed them found here, but so too were the first recognised active stromatolites, and as wide a diversity as occur anywhere in the world. Western Australia is truly a Mecca for any stromatophiles who are fascinated by these strange structures, constructed by some of the most primitive life forms ever to have evolved on this planet.

Until the stromatolites at Hamelin Pool in Shark Bay were recognised for the first time in the 1950s, stromatolites had only been known from the fossil record. Take a walk down to the beach at Hamelin Pool, preferably at twilight. There you will be met by a sight that, if your imagination is switched on, is guaranteed to send a cascade of shivers down your spine. For arising out of the sea, like serried ranks of concrete cauliflowers, and stretching as far as your eye can make out, are the stromatolites. It looks for all the world as if you had been transported back in time thousands of millions of years – to the very dawn of life itself on this planet. Time moves very slowly in W.A.

2

Rise of the Gutless Wonders

Early animal evolution

Precambrian times are characterised not just by the overwhelming dominance of microbial communities, but for the evolution of eukaryotic cells, and thus of plants and animals. Great inroads have been made during the last decade into investigating the relative timing of the appearance of more complex organisms, and their subsequent evolutionary histories. But many questions remain unanswered. Of particular interest is the question of whether the major phyla of animals appeared on the scene geologically almost instantaneously, as many have argued, or whether it was a long, drawn-out process. Herein lie two of the major challenges for the first part of the next century – discovering when animals evolved – and how quickly the major groups became established.

Intuitively we would expect the earliest fossils of eukaryotic cells, be they animals, fungi or plants, to be mere specks in the rock: just a few clusters of cells that grew together in an organised structure, and as difficult to find as a fossilised bacterium in a mountain. After all, dinosaurs started relatively small and got bigger as they evolved; so too did humans. So why not the first plants? The trouble is that with many groups of animals and plants, evolution does not necessarily proceed, as we might expect, in a nice, orderly fashion from the small and insignificant to the large and more complex. Take, for instance, the male of the living eulimid gastropod *Enteroxenos.* This tiny beast is little more than just a bag of testes, living parasitically within a female of the same species, which itself lives parasitically within a sea cucumber. Here, evolutionary success is measured not by any great acts of machismo, but by its simplicity and minute size. Yet this little naked gastropod lies at the end of one particular evolutionary path, that began more than 500 million years ago. Size had nothing to do with evolutionary 'success'. Any species that managed to evolve is a success story. So, to look for the oldest known fossil algae, do we need to peer down a powerful microscope in the hope of spotting some nondescript little speck that led ultimately to daffodils and artichokes? To find out, it is necessary to travel to, of all places, an iron ore mine in Michigan.

The spiral filaments that Tsu-Ming Han, with Bruce Runnegar of the University of California at Los Angeles, found here at the Empire Mine in Marquette, in the 2100 million-year-old Negaunee Iron Formation, turned out to be anything but tiny[1]. Yet their interpretation as the oldest known eukaryotes pushed the date back hundreds of millions of years in one fell swoop. If it were possible to stretch out these fossil filaments, they would unravel to be, surprisingly, amongst the largest of any Precambrian fossils known to date, growing up to an amazing half a metre long, although just 2 mm in diameter. Known as *Grypania*, similar spiral fossils have been found in much younger Precambrian rocks 1100 to 1400 million years old in Montana, China and India. *Grypania* had a circular cross-section and, as Han and Runnegar describe it, was a 'corkscrew-shaped, spaghetti-like organism'. Specimens found in India and China suggest that the organism maintained its corkscrew shape in life with the aid of helical filaments set within the body wall. Han and Runnegar consider that although there are no modern equivalents of *Grypania*, its complex morphology, large size and rigid structure, indicate that it was a true megascopic alga.

Whether eukaryotes evolved from Bacteria or from Archaea is open to question, although some authorities think that the evidence points most strongly to the latter. It is possible that these earliest of eukaryotic cells did not possess organelles like mitochondria and chloroplasts, because it has been suggested that their absence in living amitochondriate protists, such as *Giardia*, indicates they were also absent in early forms of eukaryotes. If this were the case, then the earliest pre-*Grypania* eukaryotes, which Han and Runnegar believe may have lived as far back as 2500 million years ago, were incapable of respiration. However, recent research suggests that in at least some eukaryotes that lack mitochondria, they were originally present, but subsequently lost. The evidence for this comes from the presence of characteristic mitochondrial genes in the nuclear chromosome that indicate transfer of the mitochondria. Han and Runnegar think that *Grypania* was capable of aerobic respiration. If so, this suggests that the atmosphere round about 2000 million years ago must have had at least 1% of the present atmospheric level of oxygen.

Other very old, macroscopic algal remains of very different appearance have been found in younger Precambrian rocks. Back in the ancient Pilbara region of Western Australia, impressions of strings of interconnected beads were found on extensive bedding surfaces of sandstone by Kath Grey and Ian Williams of the Geological Survey of Western Australia in the early 1980s[2]. These show a remarkable resemblance to some living seaweeds. But the fossil seaweeds are preserved in rocks dated at about 1300 million years old. Very similar looking structures have been found in Precambrian rocks in Montana, in rocks about 1500 million years old[3]. In all likelihood, as Precambrian times progressed the seas would have become carpeted by sweeping meadows of seaweeds. Despite the fact that seaweeds are rare as fossils in the Phanerozoic, their lack of a fossil record is no real guide to their true abundance and diversity during the Precambrian. But just when were these tranquil waters, the domain of bacteria and seaweeds, first disturbed by multicellular animals? When did these animals (known as the Metazoa) first evolve, and just what was the early history of their evolution and diversification?

Not surprisingly, the early fossil history of the first animals is spotty to say the least. Actual 'real' fossils that you can pick up, take one look at and say, 'yep, there's no doubt – that's a fossil', appear in the fossil record during the last gasp of the Precambrian, between about 600 and 540 million years ago. These fossils are so distinctive, and so unusual, that they have been called the Ediacaran assemblage, named after the Ediacara Range in South Australia. This is close to where they were first recognised in Australia by the geologist Reg Sprigg in 1946. The same types of fossils had actually been found much earlier in three other continents. Discs found in Charnwood Forest in England and described in 1840 may well represent the same structures that were realised in the 1950s to be Precambrian fossils (see p. 23). E. Billings first described a fossil that he called *Aspidella*, and which is now recognised as having affinities with the Ediacaran assemblage, in Newfoundland as long ago as 1872. This fluted, pancake-like fossil is very similar to a form called *Cyclomedusa* subsequently found in South Australia. Then, in 1933, G. Gürich described *Paramedusium*, *Pteridinium* and *Rangea* from rocks of similar age from Namibia in southern Africa. Fossils like these had in fact been noticed by German settlers before the First World War. These delicate, ribbon-like and frond-like structures were large, growing up to 60 cm long. Regional pride being what it is, Bob Brain, former head of the Transvaal Museum, has argued that the assemblage really ought to be known as the 'Naman' fauna, after the Nama sediments in which they occur. But common usage rather than precedence by age has meant that the name Ediacaran has stuck for all these strange and wondrous creatures that wafted in the world's early oceans.

Surprisingly this assemblage, which contains a range of apparently soft-bodied jellyfish-like animals, is geographically very widespread. In addition to Newfoundland, Namibian and South Australian occurrences, they have been found in more than 30 places around the world, including sites in subarctic Russia, northern Siberia, Ukraine, China, England and, most recently, Western Australia. In what some scientists think may be the geologically oldest deposits, such as those on the Avalon Peninsula in Newfoundland and northwestern Canada, the animals appear less 'complex' than younger Ediacaran assemblages. It is thought that they might have begun to diversify during a major global glaciation (known as the Varangian glaciation), evidence of which can be seen in many places, such as in northern Norway and the Mackenzie Mountains, in western Canada[4]. Odd little disc-like fossils, called 'Twitya' discs, found in northwestern Canada occur in inter-tillite beds that formed during this glaciation[5]. The maximum diversification of the Ediacaran assemblage came after the last glacial maximum. The question is: were these two factors, the increase in diversity of the assemblage, and the waxing and waning of glaciers, in any way related?

So, what do these first animals (if indeed that is what they are) look like? In many respects they are thought to have resembled submersible air mattresses. Many were jellyfish-like, as big as dinner plates, with bands of concentric ridges. Others had radiating grooves; some had both. Others had curious three-legged like structures in the middle, resembling an early design for the Isle of Man flag. Several were elongate, ribbon-like structures, while a few resembled living flatworms. It stretches the imagination to think that such a variety of organisms could have sprung forth fully formed during this very short period of geological time. Where are their ancestors? And are they the ancestors of the vast plethora of groups that gatecrashed the Cambrian? There are several lines of evidence to suggest that metazoans had a reasonably respectable pedigree, despite some arguments that the major phyla appeared very quickly over less than 10 million years at the beginning of the Cambrian. Teasing out when, and how the phyla evolved, though, is by no means an easy task.

As I mentioned in the last chapter, the rather dramatic decline in stromatolites round about 1000 million years ago is usually attributed to the rise of animals. But what sort of animals? And have they left any fossil record? Active burrowing and grazing by these mythical animals, of the microbial mats that covered the stromatolites or their affect on the water chemistry by contributing to increasing nutrient levels, are the usual factors that are said to have led to the stromatolites' decline. It is of course possible that the decline of stromatolites could have been caused principally by non-biological factors. After all, they are merely sedimentary structures, dependent on their growth for factors other than just biological input.

Before entering the debate on where the Ediacaran fossils fit into the great evolutionary scheme of things, I want to examine the evidence, however scant, for the animals that might have lurked in the ocean depths even before the Ediacaran fossils. The only direct evidence pointing to the existence of animals in pre-Ediacaran times are trace fossils: the enigmatic burrows and trails that record the activity and behaviour of extinct organisms for a few fleeting moments of their existence far back in deep time.

How old are the oldest Precambrian trace fossils, and what can they tell us about how complex and diverse were these first animals? A variety of curious structures preserved in sedimentary rocks dating back to about 1000 million years old have been attributed to the activity of the first animals, scratching and scraping their way through the sediment. However, deciding on whether such structures were of biological, rather than physical, origin can be difficult. Are the 'problematic structures' found on bedding surfaces in the Glacier National Park by the late Bob Horodyski evidence of the scrapings of some of the earliest animals[3]? If so, this would put back the origins of animals to at least 1500 million years ago.

However, recent reviews[6] agree that the prize for the earliest evidence of metazoan activity should be awarded to the trace fossil called *Planolites* found in rocks that formed somewhere between 750 and 900 million years ago in present-day Namibia[7]. Less certain is a structure that could also represent metazoan activity, interpreted as a possible backfilled burrow and tentatively referred to as *Bergaueria*. This particular trace fossil was found in 800 to 1100 million-year-old rocks in the Mackenzie Mountains in northern Canada. Another interpretation for this structure is that it may be the resting trace of a jellyfish-like organism.

While the evidence for animal activity more than 700 million years ago is flimsy to say the least, we are on firmer ground with structures preserved in rocks between 540 and 700 million years ago, and which, for the large part, predate the Ediacaran fossils. *Planolites* has also been found in the Georgina Basin in Australia, in rocks deposited a short time (geologically speaking) before the Ediacaran organisms lived. The only other reasonably definite trace fossil of this age is a sinuous, meandering trace called *Cochlichnus*, found in New South Wales in Australia[8]. Such sinuous loops and trails therefore characterise the sparse evidence that animals did indeed live and flourish before the weird world of the Ediacaran creatures. And with their arrival came a great increase in number and complexity of trace fossils. To date 35 different types have been identified in association with the Ediacaran assemblage[7] although recent work suggests that this total might be a little excessive.

There is conflicting evidence from sources other than the fossil record about exactly when animals may have evolved. This evidence is indirect, coming not from palaeontology, but from molecular biology. Greg Wray, Jeffrey Levinton and Leo Shapiro of the State University of New York at Stony Brook have argued recently on the basis of an analysis of rates of molecular sequence divergence, that the major metazoan phyla diverged more than 1000 million years ago[9]. They suggest that chordates (the group including all vertebrates) diverged from the invertebrates not in the Cambrian Period, as is usually supposed, but twice as long ago, way back in the Precambrian. The protostomes (that group including animals such as arthropods and annelids that have spiral, determinate embryonic cleavage, with an embryonic blastopore at the rear near where the future mouth will be) they think separated from chordates before echinoderms. The way Wray and his colleagues reached such a radical conclusion was to estimate divergence times based on the tendency for both nucleotide and amino acid sequences to diverge over time. While there are likely to be problems with such an approach because divergence rates are known to vary between different groups, as well as over time, Wray and his colleagues argue that when viewed in such a broad way, looking at so many groups, mean rates of divergence times can be validly used. But as I have pointed out in the previous chapter, research published early in

1998, replicating the work of Wray and his colleagues, has come up with a substantially younger age for these divergence times, about 670 million years for the divergence of the protostomes from the deuterostomes, and about 600 million years for the chordates from the echinoderms. A lot more is guaranteed to be written on this subject over the next year or so.

In their analysis, Wray and his colleagues targetted seven genes, and they investigated six phyla: Mollusca, Annelida and Arthropoda in the Protostomia, and Echinodermata, Agnatha (jawless fishes, like lampreys and hagfish) and Gnathostomata (fish with teeth and jaws) in the other major group, the Deuterostomia (these are animals that have radial, indeterminate embryonic cleavage, and have an embryonic blastopore situated where the future adult's anus will be). The protostomes and deuterostomes comprise the coelomate triploblastic metazoans (in other words, multicellular animals with a gut and three germ layers). The coelenterates (jellyfish, corals and the like) are different – they are diploblasts, having only two germ layers. Wray and his colleagues estimated that the protostomes and deuterostomes diverged perhaps as much as 1200 million years ago, very much earlier than the fossil record would suggest. Their common ancestor would obviously have been even older than this. The three protostome groups likewise diverged from each around this time. Echinoderms diverged from the chordates (the Agnatha and Gnathostomata) a little later, around 1000 million years ago. The only groups to diverge, according to their estimates, anywhere remotely near the Cambrian are the Agnatha and Gnathostomata (fish without jaws and fish with jaws, respectively), which they calculate diverged about 600 million years ago, close to the Precambrian/Cambrian boundary.

If we are to accept the findings of Wray and his colleagues, then two conclusions emerge from these findings. The first is that the triploblastic metazoan phyla may have evolved well back into the Proterozoic, a good twice as long ago as is usually argued. Secondly, the major phyla did not all burst into being around the same time in a great evolutionary explosion, which is often posited for the beginning of the Cambrian Period. Wray and his colleagues argued for a more protracted origin for these divergent groups spread over maybe as much as 600 million years, the Agnatha and the Gnathostomata being the new kids on the block.

When they looked at the more limited data base available for another 12 phyla, all indicated divergence times deep into the Precambrian. Such a long, drawn-out history for the emergence of the major phlya is very much at odds with other estimates. Some scientists have argued that the metazoan phlya burst forth over a period as short as 8 million years. For this to have happened, it would be necessary to argue that early in their evolutionary history, all phyla quite independently have extremely rapid rates of divergence in a range of different genes. The amount of change in a few million years would have to equate to all the other change that occurred over the subsequent 600 million years – a more than 60-fold difference in divergence rates. This would seem to be very unlikely.

So where does this leave the first fossil animals – the Ediacaran assemblage? If all the modern phlya were in existence at this time, all the different forms that have been found should slot quite easily into known modern phyla. Ah, if only life were that simple. And why is there no fossil record of these first animals that lived more than 1000 million years ago? Is it possible that the more recent molecular study carried out by Ayala and his colleagues is closer to the truth, meaning that the absence of animals from the fossil record between 600 and 1000 million years ago is simply because they did not exist? Or have we palaeontologists been barking up the wrong gullies in our search for the fossils? Were the early animals just too small to be fossilised? Maybe they superficially resembled today's meiofauna, tiny animals many only a millimetre or so long, that have a world all of their own, living between grains of sand (although these living forms are not primitive, but probably very derived). If bacteria can be fossilised, why not these first Lilliputian animals? Or were they living in such an active environment that upon their death their remains were destroyed too rapidly to enable them to be fossilised?

Recently, Eric Davidson and Andrew Cameron of the California Institute of Technology, along with Kevin Peterson formerly at the University of California at Los Angeles, have suggested that there was a period before the Ediacaran assemblage existed, when metazoans were microscopic, like modern-day larvae[10]. They believe this cryptic early phase in metazoan evolution left no fossil record. Their argument is that by looking at the embryological development of modern groups we can gain an insight into the evolution of metazoans as a whole. They stress that there is a tremendous difference between larvae and adults in many groups of animals. Take a sea-urchin. Its free-floating phase bears as much resemblance to a spiky sea-urchin as a gum boot does to a haggis. Where such a disparity exists they call it 'maximal undirectional development'. From this, forms with 'direct development' evolved –

shortening of the larval phase and precocious development of the juvenile form. However, there are some problems with such an argument. One of the problems is that only large metazoans produce free-living larvae, and indirect development may well not be a primitive trait. Quite probably the evolution of a free-swimming, planktic larval phase evolved independently in different groups.

Davidson and his colleagues consider that evolution of large metazoans, like the Ediacaran beasts, and the great zoological garden of creatures that seem to have sprung forth from the early Cambrian mud, arose from tiny, unknown precursors. They argue that this occurred by the evolution of a system of genetic regulation of development that produced groups of unspecified 'set-aside cells', along with an hierarchical system of regulation of development. This freed the animals from their size limitations and led to the evolution of radically different post-larval features which supplanted, in many cases, the larval structures. So not only did they discover the secret to large size, but they transformed, Jekyll and Hyde-like, overnight. The developmental shackles were released and an explosion of metazoans burst into the sea.

A recent discovery in China shows that Davidson and his colleagues were right – and that they were wrong. They were right in that exceptionally well-preserved multicellular life has been recovered from 570 million-year-old rocks in southern China[11]. They were wrong in that these microscopic embryos have been preserved. Initial investigation indicated the presence of both sponges and embryos of multicellular animals that had bilateral symmetry. Here at last appears to be confirmation that animals did exist in the late Precambrian.

But were the Ediacaran fossils part of this great evolutionary transformation? Or were they on a sidetrack all of their own? In many ways the Ediacaran fossils raise far more questions than they answer. By and large these fossilised remains are, as I have indicated, either circular, and jellyfish-like, or elongate, and frond-like or even similar to segmented worms. Some grew to large sizes, in excess of 1 metre long. After Reg Sprigg came across this wide diversity of forms in the Flinders Ranges in 1946 he believed that the discoid forms, like *Cyclomedusa*, represented jellyfish or hydrozoans. But was he correct to interpret them even as animals? In the mid-1950s Roger Mason, then a schoolboy, found the fossilised impression of a fern-like organism in Charnwood Forest in Leicestershire in England. This was subsequently named *Charnia masoni* by Trevor Ford[12] who interpreted it as an alga. So are we dealing with animals and plants, or, indeed, could we be dealing with something altogether different? In some ways many of our problems are answered if they were plants. Then we wouldn't have to worry about trying to evolve later animals from them.

More detailed work on the Ediacaran fossils from South Australia was subsequently carried out by Martin Glaessner of the University of Adelaide in South Australia in the 1960s. He argued strongly that not only were they animals (which Ford came to accept), but that many could be accommodated within living animal phyla[13]. They were not bizarre, evolutionary experiments. He even placed some in modern classes and orders of animals. Thus, he interpreted the frond-like *Rangea* and *Charnia* as pennatulacean octocorals (commonly known as sea pens); and the segmented, elongate *Dickinsonia* and *Spriggina* he compared with the living polychaete worms *Spinther* and *Tomopteris*. It is now thought likely that this similarity is more apparent than real, arising from evolutionary convergence. The generally held view by the Australian palaeontologists, Bruce Runnegar, Jim Gehling, Mary Wade and Richard Jenkins, who have worked on the Ediacaran fossils in more recent years, is that the Ediacaran assemblage was dominated by solitary and colonial cnidarians (the group containing corals and jellyfish), plus annelid worms, arthropods, and possibly echinoderms. Yet, according to Bruce Runnegar[4] despite this, Glaessner and later Australian researchers have not claimed that the Ediacaran animals were the ancestors of all later groups that burst on the scene early in the Cambrian, just a few million years later.

The groups in which they are placed are relatively specialised groups, like sea pens (*Charnia* and *Charniodiscus*), phyllodocid polychaete worms (*Spriggina* and *Dickinsonia*) and edrioasteroid echinoderms – a class that became extinct in the Palaeozoic – (*Arkarua*). Runnegar has argued that this Ediacaran assemblage of very distinctive organisms represents just a tiny part of the dying embers of the Precambrian world that lived alongside the ancestors of the phyla that appeared on the scene so dramatically at the beginning of the Cambrian.

Other people have had very different ideas about the Ediacaran fossils – ideas of quite astounding originality. Hans Pflug of the Justus-Liebig University in Giessen was so struck by the seeming uniqueness of the Ediacaran fossils that he proposed that they should be placed in their own, separate phylum, that had no living representatives. This he called the Petalonamae. He interpreted all these organisms as having lived quiet, sessile lives, feeding off plankton. What united them all into one phylum was their construction that consisted, according to Pflug, of

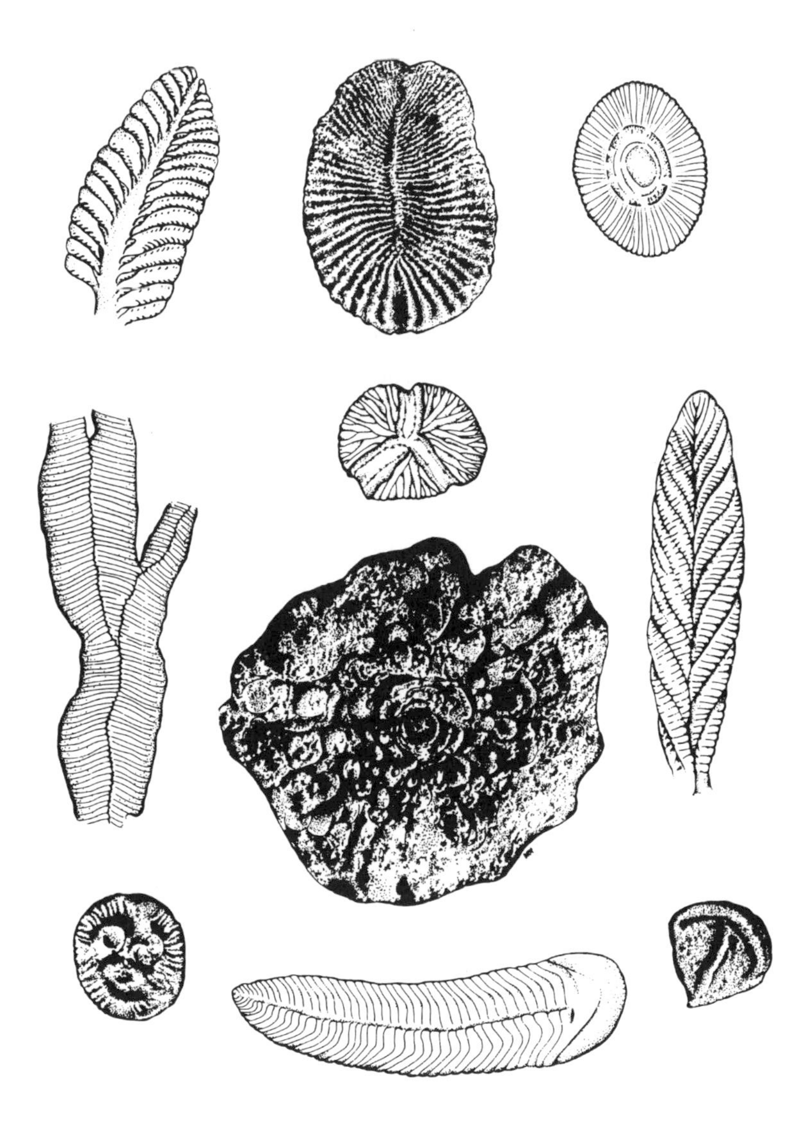

Animals, plants, or now for something completely different? These impressions of ancient life from Ediacara *in the* Flinders Ranges *in* South Australia, *ranging in size from one to a hundred centimetres, date back some 600 million years, and represent one of life's first explosions of multicellular diversity. Debate still rages over whether they are allied to living groups of animals, or are closer to plants, or belong in their own unique category of life form, the* Phylum Vendobionta. *Clockwise from bottom left:* Tribrachidium, Phyllozoon, Rangea, Dickinsonia, Cyclomedusa, Charnia, Parvancorina *and* Spriggina; *of the two in the middle, the little one is* Albumares, *the larger one* Mawsonites.

tiny, tubular individuals joined into leaf-like petals[14]. These were either folded into a cylindrical shape, or flat. Although Pflug accepted that the Ediacaran fossils were once animals, he believed that they were separate from all others. There has been little acceptance of Pflug's idea, perhaps in part because in addition to writing in impenetrable German he erected so many different taxa that are thought by most other palaeontologists simply to represent variations caused by different modes of preservation.

But the palaeontologist who really put the cat among the pigeons, taking Pflug's idea to even greater extremes, was Dolf Seilacher. He proposed, first at a Dahlem Conference in Berlin in 1983, and also that same year at the Annual Meeting of the Geological Society of America in Indianapolis, that these strange fossils had nothing whatsoever to do with later or any other phyla. They were, in Seilacher's eyes, an extinct group of organisms that were not even metazoans. As such they could have nothing to do with the subsequent evolution of all metazoan phlya. So he called them the Vendozoa – an evolutionary experiment that failed[15]. Rather than being part of one flowing river of evolution, branching in myriad directions, here was an evolutionary experiment that worked for a short time then dried up; but by its very nature it was doomed to fail. Like Runnegar and others, Seilacher saw his Vendozoa as coexisting with modern animal phyla. As to what these coexisting creatures would have been, our only indication comes from enigmatic trace fossils.

To establish these 'vendozoans' as a group separate from animals, plants and protists necessitated an appropriate name. In 1992 Seilacher proposed the kingdom Vendobionta[16]. Within this he saw three groups: 'erect elevators', such as *Charnia*; 'flat recliners', like *Dickinsonia* and 'sediment stickers', such as *Rangea.* Seilacher's view is that vendobionts were, on the whole, large, sessile, quilted, mattress-like organisms, filled with plasmodial fluid. How they fed is a big problem. They have no apparent mouth or anus, and most show no sign of a gut. But even where there is exceptional preservation evidence of a mouth and anus is generally rare. Were they photosynthetic? Were they symbionts with bacteria? Or did they absorb organic material directly through their body walls? One thing is for sure – we don't know. Some specimens of *Dickinsonia* do, however, show evidence of a gut running down the centre: an elongate, thread-like structure filled by sediment of a grain size different from that in the surrounding rock. So perhaps a few Ediacaran animals were not quite such gutless wonders after all!

Because of the large size of many of the Ediacaran fossils (for instance *Charnia* and *Dickinsonia* could reach lengths in excess of 1 metre) it has generally been assumed that they were multicellular. But this need not necessarily be the case. As Bruce Runnegar has pointed out, some living green algae can reach lengths of 20 metres, but are single-celled hydrostats (flotation sacs). It is possible that some of the Ediacaran fossils, such as *Phyllozoon*, *Ernietta* and *Pteridinium*, were more algal-like in this regard. Seilacher is of the opinion that selection pressure favoured large body size, but that this was not achieved by multicellularity. It came about, he argues, by a proportional 'quilting' of a plasmodial mass. The problem with this interpretation is that many of the Ediacaran fossils do not appear to have this 'quilted' structure.

Cellular form can perhaps be seen in the dendritic canals that are present on *Albumares*, implying that, if correctly interpreted, some of the Ediacaran fossils at least were multicellular. This, and other complex disc-like forms, such as *Tribrachidium*, have been placed in an extinct class of coelenterates, the Trilobozoa, by Mikael Fedonkin of the Palaeontological Institute in Moscow, along with some Cambrian microfossils, called anabaritids, which also have a triradiate structure[17]. If these two do, indeed, belong within the same group, then it breaks down the perceived barrier between the Precambrian Ediacaran and the post-Precambrian 'modern' fauna.

The triradiate forms, like *Tribrachidium*, also show some developmental similarities to some living groups (cnidarians and echinoderms). If they do, indeed, belong, or are closely related, to these groups, it would imply that they had a complex 'animal-like' development. Runnegar has also argued that the apparent segmentation and strong bilateral symmetry in *Dickinsonia* and *Spriggina* may have its underlying formation in Hox gene systems that control anterior to posterior development in a wide range of animals, from mice to fruit flies. Hox genes are critical 'switch' genes that control development, because they trigger the genetic pathway that establishes the identity of body segments in the embryo. However, Seilacher would counter this argument by saying that the apparent segments of *Dickinsonia* are extra quilts that the vendobiont lays down as it grows. The trouble with this is that, as Jim Gehling of the University of South Australia has pointed out to me, if these organisms are quilted, why do the quilt elements get smaller at the 'posterior' end. Or, then again, is it a case of the elements getting bigger as the body grows at the other end? This may well be the situation in *Phyllozoon*.

Simon Conway Morris has shown that Ediacaran-type forms persisted well into the Cambrian. He sees similarities between *Charniodiscus* and a form called *Thaumaptilon* from the Middle Cambrian Burgess Shale from British Columbia (see Chapter 3). *Thaumaptilon* consists of a broad frond, with a central 'stem' and an elongate holdfast, closely resembling forms like *Charnia* and *Charniodiscus*, the sea-pen-like fossils from the Ediacaran. Conway Morris is of the opinion that *Charniodiscus* may well be a true cnidarian, and that many other Ediacaran species were diploblasts, similar to cnidarians and ctenophores[18]. Other links from the Ediacaran to Cambrian assemblages are *Arkarua* and the edrioasteroids; *Parvancorina* and *Skania*; and *Ovatoscutum* and the chondrophore *Plectodicus*.

Such apparent close similarities between some of the Ediacaran fossils and 'modern' phyla are a challenge for the view that the Ediacaran assemblage is something quite unique. But hovering in the background there is always the omnipresent spectre of convergent evolution. Can a sea pen separated in time from a fossil by nearly 600 million years, really share a heritage with it? Or do the limits on shape and size within which organisms are constrained mean that every so often very similar looking creatures are tossed up by the tides of evolution? The difficulty with this line of argument is that there are an awful lot of examples of convergent evolution occurring in the one assemblage.

Herein lies one of palaeontology's greatest challenges as we march into the twenty-first century, understanding not only the early evolution of animals, but also how the Ediacaran fossils fit into models such as those proposed by Greg Wray and his colleagues. The Ediacaran fossils include forms that are very reminiscent of living animals. The segment-like *Dickinsonia* and *Spriggina* look like coelomate triploblasts, in particular arthropod or segmented worm-like ones. *Arkarua* is reminiscent of an echinoderm; the kite-shaped *Parvancorina* may also be an arthropod; *Inaria*, perhaps an actinian coral; *Charniodiscus* and *Charnia* may be sea pens or pennatulaceans; and *Kimberella* may be a precursor of the molluscs[19]. It is relatively easy to conjure up resemblances, fanciful or otherwise, of these fossils to living equivalents. After all, that is what our training as palaeontologists tells us – the present is the key to the past.

Are they related to living animals, or an entirely extinct group? Or is there another alternative? Greg Retallack of the University of Oregon thinks there is. One aspect of Ediacaran fossils that has intrigued many palaeontologists, and which Seilacher has pointed out might just provide an insight into the biological affinities of these gutless wonders, is the manner in which they are preserved. We call them 'soft-bodied', yet they are preserved in a way quite unlike any other post-Cambrian soft-bodied fossils. Rather than being flat impressions in the rock, they often have quite prominent relief. To Greg Retallack this suggests that they were so tough that they were capable of resisting compaction to the same extent, amazingly, as fossil tree trunks. This, Retallack believes, puts paid to the notion that these organisms were in any way related to worms or jellyfish.

Rather than a comparison with animals, he points out that in terms of how they are preserved, along with certain morphological features, the Ediacaran fossils compare more closely with fungi, algae and, more particularly, a combination of the two – lichens[20]. Lichens are symbiotic partnerships between, principally, fungi with either an algal colony or cyanobacteria, or sometimes both. The latter supply the organism with nutrients by photosynthesis, while the fungal component provides the structure.

The apparent toughness of these organisms indicates to Retallack that they cannot have been soft-bodied, nor could they have been the mattress-like organisms that Seilacher proposed. However, those palaeontologists with a long history of working on Ediacaran fossils are very skeptical of Retallack's claims. Jim Gehling has indicated to me that there are many problems with the lichen hypothesis. He feels that the relief of the preserved moulds is not a reliable measure of the toughness of the original body. Retallack's analysis of just four specimens was an attempt to compare fossil mould relief with fossil wood compressions of similar burial depth, and from that to draw conclusions on the toughness of the original materials. Gehling argues that the fossils he has examined show the highest relief specimens come from the site where the depth of burial was the greatest – more pressure, more relief. In regions like the Flinders Ranges specimens of all sizes are relatively flattened, even though the depth of burial in this basin margin was probably little more than 200–300 metres. The relief we see in fossils, Gehling contends, is more likely to be a product of how quickly the organisms were fossilised, rather than just burial depth or how tough their skins were.

According to Retallack the fungal hyphae present in lichens are strengthened by structural chitin filaments, giving great strength to the organism. In support of his case Retallack points out that modern lichens show a wide range of body forms, that, with a few mental gymnastics, look vaguely like some Ediacaran fossils. The rather curious large size of the Ediacaran fossils is more easily explained, in Retallack's view, by them being lichens, because sessile, photosynthetic organisms tend to have a large body size. The meshwork of tubular cells that Pflug described in some of the Namibian fossils Retallack thinks might be the cyanobacterial or algal symbionts of the lichen. Jim Gehling, on the other hand, says that they are a mineralogical growth, where the mineral limonite has replaced pyrite!

But lichen in the sea? Surely they just hang around rocks and old trees? Not so, according to Retallack. He states that there is a great diversity of lichens today that also live in the sea, in lakes and in streams. However, today macrolichen hardly make up a large part of the marine ecosystem. One major problem for Retallack's interpretation is that some of the Ediacaran deposits appear to have been formed in deep water, that was probably below the photic zone. But then again some workers have raised doubts about these deposits as being deep water. Lots of claims and counter claims.

So, was the long Age of Bacteria overtaken (though not displaced) by a brief Age of Lichens, before animals exerted their influence so profoundly on the marine ecosystems a little over 500 million years ago? And where did the Ediacaran organisms go? Did any leave later descendants? Are you and I their far off descendants? Or were they an evolutionary experiment that 'failed' because they couldn't keep up with environmental changes that made them all too readily redundant? Or, as a number of scientists have suggested, did the rise of predators see to their demise? Perhaps, whether they were animals, plants or lichens, they were just too tasty for their own good.

3

Cambrian Dreamtime Menagerie

Hallucigenia, Jianfengia and other taxonomic nightmares

If you take a stroll down a road in Nanjing, China, called Chi-Ming-Ssu, and look up at the sky as you walk, you could be forgiven for thinking that you had been whisked back in time to the Mesozoic. OK, I know that rather than herds of busy hypsilophodontid dinosaurs racing past you, it would now be herds of cyclists. But the beautiful ginkgo and metasequoia trees that line the street and make a wonderful archway as they stretch across the road are living fossils. Both formed an important part of Mesozoic floras before the rise of the flowering plants.

It is appropriate that this road should be graced by trees with such a rich heritage, for set back from the road, in beautiful gardens of its own, is the Nanjing Institute of Geology and Palaeontology. Spear-headed by eminent trilobite palaeontologist Zhang Wentang, the Institute has for many years studied Early Cambrian rocks and their faunas in various parts of China. One area that has been concentrated on has been the region south of Kunming in Yunnan Province. In the region around Dianchi and Fuxian Lakes are rocks of the Meishucunian Stage containing the earliest rocks of the Cambrian, that once vied for the honour of being named the global type section for the boundary between the Precambrian and the Cambrian, dated at about 540 million years old.

Long known as a rich source of fossil trilobites, these rocks have been catapulted into stardom since the discovery near the town of Chengjiang of an amazing fauna of soft-bodied animals. It is rare enough for animals that lack hard parts to be fossilised, but for this Chengjiang fauna, one of the few faunas in which this has happened, to be amongst the earliest known, has been to palaeontologists rather like what a vat full of cream would be to a kitten. Every new find is lapped up with relish, and they return for more and more in the hope of never being satiated. What makes the Chengjiang fauna so special is that it existed shortly after the first appearance of animals with hard parts, marking the onset of the so-called Cambrian Explosion, and it is the oldest known soft-bodied fauna from post-Precambrian strata.

Tracking down who was the first palaeontologist to discover this fauna is an interesting exercise. Certainly, one of the leading lights in finding and describing much of the arthropod material that has come out of these fine-grained rocks is one of Zhang Wentang's colleagues at the Institute in Nanjing, Hou Xianguang. Hou made his first discovery of soft-bodied fossils on 1 July 1984, 5 km east of Chengjiang, below a hill called Maotianshan[1]. For some years Hou had been studying small arthropods called bradoriids, when he made his discovery in the Yu'anshan Member of the Heilinpu Formation (formerly known as the Chiungchussu Formation). Another to have been collecting at this time was Shu Degon of the Northwest University in Xi'an. But in fact fossils with soft parts had been found earlier than this. Luo Huilin of the Yunnan Institute of Geological Science in Kunming, who has carried out a lot of work on the stratigraphy of these Early Cambrian rocks, discovered a fauna in February 1979 of large, bivalved arthropods, called *Isoxys*, as well as some worms, called *Sabellidites*[2]. But an even earlier discovery was made by Zhang Wentang himself when in 1972 he found fossilised arthropods with some of their soft parts preserved, hinting at the rich treasures that were to come. It also seems likely that as far back as 1912 an early publication by Henri Mansuy of Chinese fossils illustrated a fossil that may well have been from this fauna.

Following the excavations that followed Hou's discovery of a rich fauna dominated by arthropods and arthropod-like animals, more than 70 species have been recognised. Chen Junyuan and others from the Nanjing Institute have since collected more than 16 000 specimens from Maotianshan. Hou realised the significance of his discovery when he identified the remains of an arthropod called *Naraoia*. At that time it was only known from the Burgess Shale in British Columbia and from

Cambrian strata in Utah and Idaho. Prior to the recognition of the Chengjiang fauna our only window into this ancient lost marine world was the Burgess Shale[3], a unit of Middle Cambrian age, about 10 million years younger than the Heilinpu Formation. Any Cambrian fossil fauna that has soft-part preservation will always be compared against the fossils of the Burgess Shale, for these rocks contain one of the most stunning fossil faunas known from any time, from anywhere in the world. So much has been written about the fossils of the Burgess Shale, I won't say *ad nauseum*, but certainly in great depth and repeated many times, that it would be superfluous to describe yet again this menagerie of strange and wondrous beasts that swam and crawled their ways through the Cambrian seas. Rather, I will concentrate on the Chengjiang fauna.

However, I will make a couple of observations, relating to the naming of some of the Burgess Shale fossils, and the recognition of their significance, as these have a bearing on the Chengjiang fauna. Back in the early 1970s the Burgess Shale and its fauna were known to just a few palaeontological aficionados. But due to the magnificent efforts of Harry Whittington, Simon Conway Morris and Derek Briggs working at the University of Cambridge, aided by the popularisation of the fossils in the 1990s by Stephen Jay Gould in his book *Wonderful Life*, the fauna now has achieved almost cult status. But there were a few other players in the Burgess Shale saga. And I reckon that I must hold the record for having played the smallest, and most insignificant part possible.

I undertook research for my PhD at the University of Cambridge at the same time as Simon and Derek, and like them I was supervised by Harry Whittington. My research topic was on more prosaic matters, Late Ordovician trilobites from northern England. Having chosen the seemingly more obscure and less sexy part of the fauna – the worms and other wierdos – Simon, who undoubtably knew what he was doing, had the happy fortune to be burdened with many new taxa, including both undescribed species and genera. Coming up with names for new fossils is often one of the more delightful aspects of being a taxonomist. In the case of some of the bizarre Burgess Shale fossils, it was possible to let the imagination run riot, and this we often did on the long bus ride home after an evening sampling the delights of the best East Anglian brewers. One of the better efforts was Simon's name *Hallucigenia*, for one of the strangest creatures from this fauna, but which, with further discoveries from Chengjiang, has turned out not to be as strange as first thought. My great claim to fame was the suggestion of the name of one of the worms. Its striking resemblance to a bottle brush seemed too good not to go unnoticed. Simon latinicised 'bottle brush', and hey presto *Lecythioscopa* was born.

My own introduction to this amazing fauna came when I acted, I suppose, as a sounding board for many of Simon's early ideas. For reasons that are not worth recounting, neither Simon nor myself partook of one of the great Sedgwick traditions, morning and afternoon tea on the ground floor. Our establishment of a 'splinter tea room' (and I use the word 'room' in the broadest possible sense) didn't go down too well with some of the more long-standing members of the department, but for three years we persisted in somehow managing to make not only tea, but also cook crumpets on an old sediment drier in Simon's apology for a room. When I say room, it was really little more than a cupboard at the farthestmost corner of the top floor of the Sedgwick Museum. There was just enough room for the trays of fossils, Simon, me and a kettle. As he slowly prepared out the fossils, so the ideas on what was emerging from his dental drill slowly, and sometimes not so slowly, emerged.

Two fossils that stand out, perhaps because they are amongst the two that have engendered most debate since, were *Hallucigenia* and *Pikaia.* And the Chengjiang fauna, as I will show, has since helped clarify the status of both of these two strange fossils. The careful excavation of *Hallucigenia,* that appeared as a tube with paired spines on one side and what seemed like a row of tentacles along the other side, caused Simon the most problems. The 'tentacles' seemed so flexible that all he could suggest they were used for was the transfer of food to the mouth (though which end that was at was another problem). The paired spines therefore had to be the 'legs', though how it could locomote on such stilts was hard to imagine.

Much has since been written on how Simon got it 'wrong'. Fossils from Chengjiang studied by Lars Ramsköld, then of the Swedish Museum of Natural History, and Hou Xianguang revealed that Simon's reconstruction was the wrong way up[4]. But based on the material to hand at the time it seemed the most likely explanation. The benefit of hindsight is always a most useful commodity. In 1989 two specimens of a caterpillar-like animal, 5 to 6 cm long, and bearing 11 pairs of legs, were found at Chengjiang. This animal, a lobopodian, since called *Onychodictyon*, possessed paired, stubby legs with tiny claws[5]. Each of the leg-bearing segments carried a short spine emerging from a stout plate. These plates were comparable to the plates on another Chengjiang lobopod called *Microdictyon.* Such plates had long been known from Early Cambrian deposits, but their affinities were unknown. However, discovery of complete specimens of *Microdictyon* showed these plates to form as pairs along the back of the animal's trunk-like body[4]. The spinose plates of *Onychodictyon* were comparable. By flipping *Hallucigenia* over, having spines coming off the back and the 'tentacles' as legs, suddenly, rather than being a strange animal that looked as if it came from another planet, we had an animal that could be slotted into a group with other fossils. Moreover, when Ramsköld further prepared the holotype of *Hallucigenia sparsa* from the Burgess Shale, he uncovered evidence that the apparent single row of 'tentacles' were in fact paired structures[6]. These three animals, along with other similar forms, *Xenusion* from the Early Cambrian of Sweden, and *Luolishania* and *Paucipodia* from Chengjiang, were recognised as belonging in a group of lobopods called the Onychophora, or velvet worms, by Ramsköld and Hou. *Hallucigenia* is also known from Chengjiang. Onychophora, looking like a cross between a caterpillar and a worm, still thrive today, but on land, rather than in the sea.

An even more odd-looking lobopod-like animal was described by Graham Budd then at the University of Cambridge from the Early Cambrian Sirius Passet fauna in north Greenland[7]. This is another so-called 'soft-bodied' fauna, but one which has some animals quite unlike those from comparable faunas in other parts of the world. The lobopodian that Budd described, which he called *Kerygmachela*, is notable in possessing a gill-like structure, attached to lateral lobes that run along the side of the body, together with the stubby legs characteristic of lobopods. Budd has argued that the presence of the gills indicates that *Kerygmachela* may represent an intermediate link between more generalised lobopodians and biramous-limbed arthropods. In this regard *Kerygmachela* is a little like some of the more unusual Burgess Shale arthropods, such as the large, predatory *Anomolacaris*[8], and the five-eyed, nozzle-nosed *Opabinia*[9].

Of particular significance to Ramsköld and Hou's reinterpretation of *Hallucigenia* is what it means to our understanding of the nature of the Cambrian Explosion. Faunas like those at Chengjiang and in the Burgess Shale contain a wealth of organisms showing a wide range of body plans. This has led some people to suggest that a number which do not fit neatly into any modern higher taxonomic groups may well represent phyla that are now extinct – evolutionary experiments that

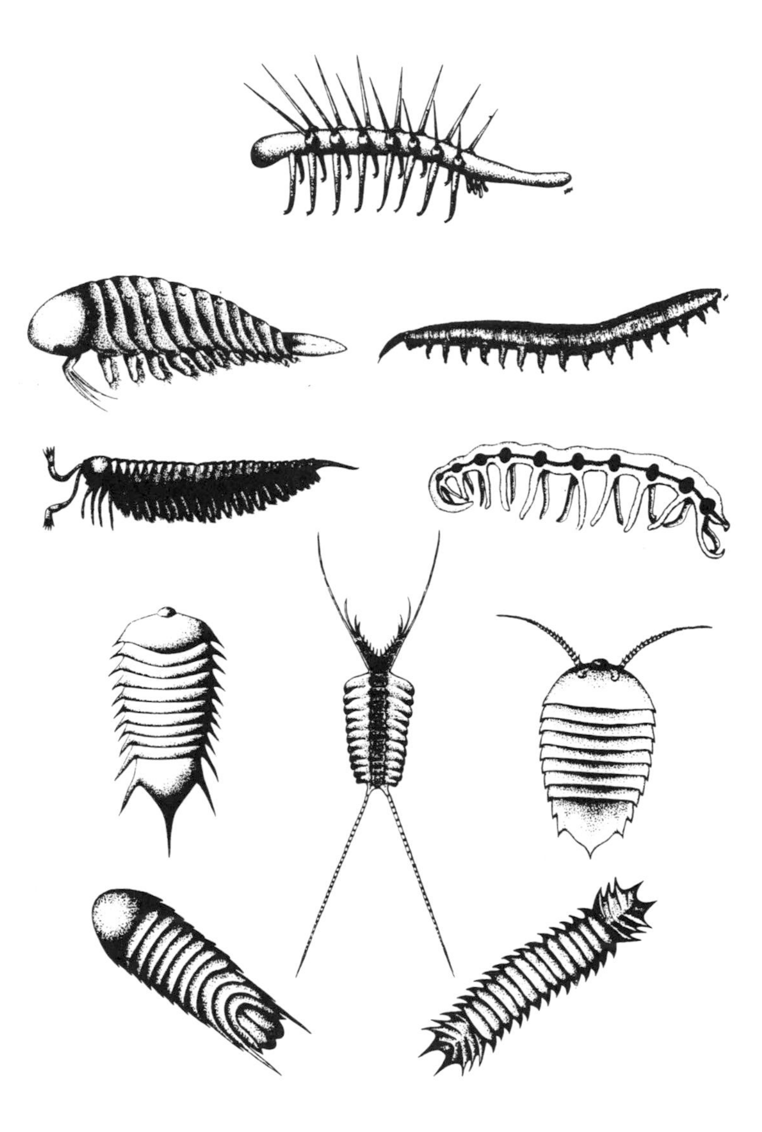

And then there were legs! The world's most primitive arthropods, onychophorans and other closely related creatures all, except one, from 530-million-year-old Chengjiang fauna of China. Clockwise from bottom left: the 'arthropods' Acanthomeridion, Rhombicalvaria, Jianfengia *and* Alalcomenaeus; *the onychophorans* Hallucigenia, Peripatus *(this is the living form), and* Microdictyon; *more arthropods;* Kuamaia, Urokodia, *and the spiny beast in the centre,* Kerygmachela *from Greenland. Each of these would fit on to the palm of your hand.*

failed. *Hallucigenia* has been held up as being one such example. When interpreted as an animal that walked on stilts it attracted, as Lars Ramsköld has pointed out, quite a range of epithets, such as bizarre, dream-like, enigmatic, extraordinary, ludicrous, misfit, odd, oddball, peculiar, puzzling, remarkable, really weird, strange, wondrous and bizarre nightmare!

However, with the inversion of the beast so that its legs are on the ground and not waving around in the passing currents, all these descriptors vanished with a puff of smoke. It suddenly became possible to slot even this seemingly bizarre organism into a modern phylum, in this case the Onychophora, a group whose modern descendants wander the hills not far from where I live in Perth, Western Australia. As Simon Conway Morris has argued 'Is it realistic to talk of a multiplicity of body plans in the Cambrian, far exceeding that of the present day? . . . the argument that extant phyla maintain their identity back to the Cambrian offers no more than a tautology. Such phyla persist because the only way they can be recognised is by reference to themselves.'[10] However, the range of morphological diversity displayed by these fossils, while impressive, is no more than the range of variation seen within phyla today.

What is stunning, though, is just how diverse these very first Phanerozoic organisms are, given that apart from the strange Ediacaran fauna (see Chapter 2) these are the first animals known in the fossil record. Yet they display a range of anatomical complexities similar, in many respects, to organisms today. They undoubtably had an evolutionary history of some magnitude. But how far back does it extend? Many have argued that all bar one of the phyla appeared within the first 5 to 10 million years of the Cambrian. However, indications from 'molecular clocks' point to a much more protracted evolutionary history.

Studying the fossils from Chengjiang and from the Burgess Shale tells us more about groups like the onychophorans that dwelt in these Early Cambrian seas than just their diversity. Studying the nature of sediments in which they occur, and the associated faunas, can provide insights into aspects of the ecology of these animals, allowing us to build up a detailed picture of the environment in which these animals lived, and how they may have interacted with each other. Certain associations of fossils are quite apparent. For instance, the only four known specimens of the onychophoran *Paucipodia* occur in thin mud beds 1 cm thick that also contain abundant specimens of the medusoid-like

Eldonia[11]. This organism also occurs not infrequently in the Burgess Shale. Work by Duncan Friend at the University of Cambridge in the early 1990s has shown that, despite its apparent jellyfish-like appearance, *Eldonia* was actually an echinoderm, probably related to the holothurians (sea-cucumbers). Chen from the Nanjing Institute and his colleagues Zhou Guiqing and Lars Ramsköld, have suggested that such mud beds were mobile clouds of sediment. The *Eldonia* are found fossilised on the top of these thin mud beds. Being a floating organism indicates that the cloud of sediment was dumped on the unsuspecting onychophorans as they rummaged in the mud, the *Eldonia* having been carried along in the cloud, and settling out to the top.

Examples are known of juveniles of another onychophoran, *Microdictyon*, preserved on *Eldonia* specimens. Chen and his colleagues have argued that this could mean that these juveniles led a pseudopelagic lifestyle, joy-riding on the backs of the *Eldonia*. Another explanation could be that they were feeding on dead *Eldonia* carcasses. The long, tapering head of *Paucipodia*, they suggest, could have had mouthparts that were adapted for sucking, perhaps from organisms like *Eldonia*.

As well as *Hallucigenia*, the other Burgess Shale fossil that really stands out in my memory of tea and Burgess Shale fossils at Cambridge was a relatively rare creature called *Pikaia* (see illustration on p. 52). It was one of those instances when the penny suddenly drops, all falls into place, the light flashes on in the brain and it all seems so obvious. Here was this worm-like creature, that had been known since Charles Doolittle Walcott first described the fauna of the Burgess Shale in the early part of the twentieth century. It was, apparently, yet another of the countless worm-like creatures that occurred in this fauna. But one day Simon made the great intellectual leap. I've no idea what he actually said, but I recall him raving on about how this thing had chevron-shaped segments, just like vertebrate musculature, and that the thread running down one side just had to be a notochord. It therefore had to be vertebrate, or at least a protovertebrate. I'm not too sure that I really appreciated the momentousness of the occasion. I suspect that my reply was something along the lines of, 'Yes, very nice Simon, but do hurry up and butter the crumpet – it's getting cold'.

But like *Hallucigenia*, recently collected fossils from the Chengjiang fauna have shed more light on the early evolution of chordates and helped cement the place of forms like *Pikaia* in the early stages of chordate evolution. Considering that the Burgess Shale and Chengjiang faunas were displaced in time by at least 10 million years, as well as being on opposite sides of the world, even during the Cambrian, the two faunas are very similar in many respects. So that as well as similar onychophoran faunas, there were worm-like creatures that seem to have possessed that all-important chordate structure – a notochord – in both deposits. The trouble with the Chengjiang fauna, however, is that there has been a plethora of riches, and two contenders for the first cousins of *Pikaia* have sprung out of the pack.

In October 1995, a team, including Chen, Ramsköld and Zhou, along with Greg Edgecombe from the Australian Museum and Jerzy Dzik from the Palaeobiological Institute in Warsaw, published an article in which they interpreted one of the Chengjiang organisms, *Yunnanozoon lividum*, as a chordate. The first sentence in this article reads: 'The first chordate recorded from the Early Cambrian is the cephalochordate *Yunnanozoon lividum* from the 525 million-year-old Chengjiang fauna'[12].

They interpreted a relatively broad thread running down the axis of the body of this fat, 25 to 40 mm long worm-like animal, as a notochord. Annulations across the body they saw as being segmented musculature. This, they argued, was 'the most compelling evidence for euchordate affinities'. The other anatomical feature that led them to believe they were dealing with a chordate were structures that they interpreted as metameric branchial arches. But the life of *Yunnanozoon* as a chordate evaporated almost before the ink was dry on the paper, for in a note added in proof they point out that their 'attention has recently been drawn to the existence of an alternative possible interpretation of this species . . .'. Aware that a manuscript to this effect was in the offing, Chen and his colleagues were quick to point out that this alternative interpretation was 'unlikely'.

Six months later and *Nature* published the 'alternative possible explanation'. This was written by D. Shu, X. Zhang, and another Chen, this one, L. Chen, all from the Northwest University in Xi'an in China. Shu had been acknowledged in the paper by the Nanjing Chen and co., following a public debate on the affinities of *Yunnanozoon* that was held at a meeting on the Cambrian Explosion at Nanjing in April 1995. Shu and his colleagues had argued that there was another way of interpreting *Yunnanozoon*. Rather than being a full-blown chordate, they suggested that it was a hemichordate – worm-like animals with some chordate affinities, as shown by the presence of pharyngeal slits and a hollow, dorsal nerve cord[13]. As such it possessed 'half of the characteristic chordate features and providing an anatomical link between invertebrates and chordates'. Today hemichordates are represented by enteropneusts (acorn worms) and pterobranchs. The enteropneusts are worm-like creatures that are mobile and burrow in mud. Only 11 genera are known today. They are characterised by a tripartite division of the body into the proboscis at the front; a short collar, known as a mesosome; and an elongated trunk (the metastome).

While the Nanjing team had 25 specimens of *Yunnanozoon* to work on, the Xi'an team had more than 40, including both juveniles and adults. What these appear to show are the characteristic hemichordate features of proboscis, collar and trunk. Rather than being segmented musculature, the Xi'an team argue that it is a 'peculiar sclerotised dorsal fin'. *Yunnanozoon* lacks the characteristic chevron-shaped pattern of chordate myomeres, such as occur on *Pikaia*, and is set only in a dorsal position. They likewise argue that the 'notochord' cannot be so because it is curved, and seemingly not stiff, as it should be. More likely, they argue, it is the gut as it appears to contain gut contents. However, some palaeontologists are unconvinced by this hemichordate interpretation.

So, the Chengjiang chordate lived for just six months. But seven months later, phoenix-like, it arose from the ashes in quite another form, in the guise of yet another new taxon from this amazing fauna, called *Cathaymyrus*. Two of the Xi'an team, Shu and Zhang, joined this time by Simon Conway Morris, described a single specimen, *Cathaymyrus diadexus* that 'has a striking resemblance' to *Pikaia gracilens*[14]. Few would argue against this being a primitive chordate. Like *Pikaia*, *Cathaymyrus* has an anterior pharynx, which seemed to have been quite large, myotomes, and a possible notochord. While *Cathaymyrus* seems to have gill slits, these have not been reported in *Pikaia*. This 22 mm long, eel-like creature is the only one so far found in the more than 16 000 fossils collected from the Chengjiang deposit.

It is interesting that in both *Pikaia* and *Cathaymyrus*, the notochord does not extend the entire length of the body, being present only in the posterior half. As Shu, Conway Morris and Zhang point out, in the living primitive cephalochordate *Branchiostoma*, the notochord only extends into the anterior part of the body later in development. The overall similarity between *Branchiostoma*, *Pikaia* and *Cathaymyrus*, has led them to suggest that it would not be unreasonable to include these two fossils in the same group, acknowledging that there are some differences. Moreover, the similarity between the Chengjiang and Burgess faunas is reinforced, as is the similarity between elements of these most ancient of faunas and some living organisms.

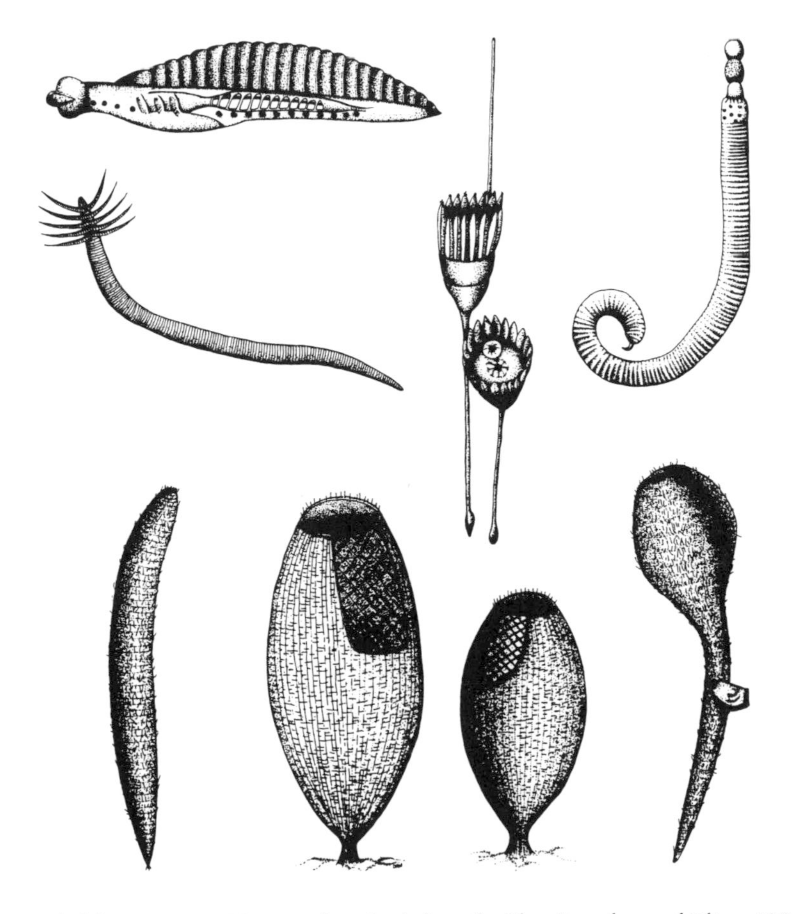

Life's wonderful experiments. Bizarre early animals from the Chengjiang fauna of China, 530 million years old. Worms, wormy-like creatures and filter feeders. Clockwise from bottom left: the sponge Paraleptomitella, *the worm* Facivermis, *the enigmatic chordate or hemichordate* Yunnanozoon, *the possible crinoid* Dinomischus, *the worm* Maotianshania, *and the sponges, another species of* Paraleptomitella *and two species of* Quadrolamoniella. *In terms of their size, each would fit on to the palm of your hand.*

Please do not think, for all my concentration on these slightly unusual Chengjiang organisms, that they dominate the fauna. They do not. The Chengjiang, like the Burgess and the Sirius Passet faunas, was a world of sponges and arthropods. Sponges probably got something of a head start in the race for colonising the world. Recent investigations by Martin Brasier of Oxford University and colleagues of microscopic remains from rocks in southwest Mongolia that are thought to be equivalent in age to the Ediacaran fauna (see Chapter 2), have yielded sponge spicules[15]. Moreover, sponges have been identified in the Ediacaran faunas in South Australia[16]. These are derived from hexactinellid sponges, a group previously not known from rocks older than Late Cambrian in age. These Mongolian spicules represent the oldest remains that can unquestioningly be assigned to an extant phylum. Moreover, they provide the first evidence for the existence of filter-feeding organisms and the presence of metazoans with siliceous biomineralisation, *before* the beginning of the Cambrian.

Many of the Chengjiang sponges are very well preserved, and although more than 20 species have been recognised, few have been formally described. In many respects they are very much like the sponges that occur in the Burgess Shale, providing another indication of the similarity between these two faunas. At least six sponge genera are common to both faunas. The other major group, one that dominates the fauna, is the Arthropoda. Some 46% of the Chengjiang fauna are arthropods. This compares with 43% in the Burgess Shale. As well as similarities in proportions to the Burgess Shale, the actual types of arthropods are very similar, even at the generic level. For instance *Naraoia*, a very common element of the Chengjiang fauna, is also present in the Burgess Shale, as are about half a dozen other forms. The Chengjiang arthropods fall into three groups. These are the 'crustacean-like' forms, the 'trilobitomorphs', and then there are those that look unlike any of today's living arthropods. A classic example of this is *Jianfengia*. This, and another arthropod called *Alalcomenaeus*, both described by Hou Xianguang, are what have been termed 'great appendage arthropods', for out of the front of the head emerges a pair of enormous jointed 'appendages' that terminate in a cluster of spines[17]. The body of *Jianfengia* has 22 segments, beneath which hang not legs, but overlapping, fringed lobes. In all likelihood it used these lobes like paddles for sculling through the water. In possessing such lobes, *Jianfengia* and *Alalcomenaeus* (also known from the Burgess Shale) are like a number of rather strange Burgess Shale arthropods, such as *Yohoia*, *Opabinia* and *Anomalocaris*, as well as *Kerygmachela* from the Sirius Passet fauna.

It is hard to know whether apparent similarities between some of these Early Cambrian arthropods and living or later forms represent valid evolutionary relationships or not. One of the Chengjiang arthropods that is hard to compare with any living form is *Fuxianhuia*, a multisegmented beast with more than 30 body segments set behind a short, broad head[18]. In tapering at the back there is a strong resemblance to some euthycarcinoids, a pretty obscure group of extinct arthropods that, as I discuss in Chapter 7, may have played a key role in the evolution of insects. Like euthycarcinoids, *Fuxianhuia* had a thorax with many appendages, behind which followed a limbless abdomen. However, euthycarcinoids have uniramous appendages (one per segment), whereas *Fuxianhuia* is biramous (two per segment). But with no intervening taxa between the Early Cambrian and the earliest euthycarcinoid, some 100 million years later, who is to say that the superficial resemblance between them is not just a function of the limit of the range of body plans that development can construct?

However, what is clear is that the close similarity between the Chengjiang and Burgess Shale fossils reflects the existence of a fauna, dominated by arthropods and sponges, that was widespread both in space and through time[19]. As well as the Chengjiang, Burgess Shale and Sirius Passet faunas, similar Early Cambrian 'soft-bodied' faunas have been found in Australia, Poland, Spain and Siberia, further emphasising their widespread nature. The Australian fauna, in the Emu Bay Shale of Kangaroo Island, is much more restricted in diversity, but it does share in common with both Chengjiang and the Burgess Shale, the huge *Anomalocaris*, the top-line predator of the early Cambrian seas. Reaching up to 1 metre in length, this multisegmented killing machine was equipped with a jaw mechanism looking like a slice of pineapple armed with teeth, and a pair of huge, segmented claws.

Studies of predation on trilobites by Loren Babcock of Ohio State University have shown that many of the healed scars on the trilobites' bodies match the shape of the mouth parts of *Anomalocaris*[20]. A startling outcome of Babcock's study was the revelation that the trilobite predators, presumably including *Anomalocaris*, show evidence of having exhibited strong right–left behavioural asymmetry. Predation scars on the trilobites reveal 70% to have been preferentially attacked on the right side. Many animals today show this same phenomenon, which arises from lateralised nervous systems, indicating that such systems were present in arthropods as far back as Early Cambrian times.

The not infrequent occurrence of these spectacularly preserved Early and Middle Cambrian faunas begs the question of why similar soft-bodied preservation in the marine realm was so exceedingly rare after Cambrian times. A number of reasons have been proposed[5]. Sedimentological data indicate a lack of bioturbation, suggesting little activity by burrowing organisms. Indeed, only a small number of the fossils present in these faunas can be inferred to have lived an infaunal existence. The chances of fossilisation would be enhanced without scavengers and bioturbators. The preservation of soft, or least weakly sclerotised, tissue indicates a lack of decay. It is possible that this could have been due to uniformly lower oxygen levels at this time, atmospheric oxygen levels probably having been lower than today. So rather than extraordinary circumstances having caused these fossils to be preserved, it was more likely the absence of destructive agents at work after the death of organisms. But whatever the cause, these new discoveries are opening the windows that allow us to take a peek into some of the most ancient life in the world's earliest seas.

4

Head or Tails?

The incredibly lucky rise of the first vertebrates

After a gruelling 2000 km, five-day four-wheel drive journey from Perth along the infamous Gunbarrel Highway, I was almost at Alice Springs, in the dead centre of Australia. Here I was, finally, at 3 o'clock in the afternoon, standing on top of Mt Watt, the site of the world's oldest known fossil fishes. I wondered whether this locality would remain as one of the oldest indisputable vertebrate sites for much longer, or would someone else find in another place older and maybe even better preserved early fishes?

The sky was an ominous dark grey as black clouds began cascading over the mountains. As I furiously swung my hammer at the 480-million-year-old buff-coloured sandstones, the fading light reflected off occasional patches of regularly patterned impressions in the rock where fish bones were once entombed. Finding a good specimen, the outline of part of an armoured plate could be made out. Right here in my hand was a piece of *Arandaspis*, a jawless, primitive fish that once swam in a shallow inland sea in the heart of ancient Australia, at a time when continents had no recognisable political or geographic boundaries. In deep time, and from a formless, undefinable land, the evolution of the first fishes seems to have happened like a flash of lightning from the sky of life.

As I pondered this miraculous step in the history of life, I felt the first heavy drops of rain plop on my shoulders. Within minutes I had to scramble down the scree to my tent. The storm blew with such intense ferocity that later that evening I had to abandon my torn and shapeless tent and make a dash for the back of the Landcruiser, along with my two colleagues. We spent a very uncomfortable night there, sleeping hunched up in the vehicle on top of our supplies, while fierce winds and thick rain lashed the desert landscape around us. Still, I remember seeing the sunrise next morning, the first rays beaming down on Mt Watt, which, bathed in primal light, appeared to me like a monument to that great evolutionary step, the rise of the first fishes.

To many of us the word 'fish' conjures up a culinary vision of a delicious dinner to be served with a slightly dry, chilled white wine. Nonetheless, we all think of living fishes as simple creatures with a bony skeleton that swim, have fins and gills, and reproduce by laying eggs externally. But nearly all of these characters can be found in creatures that are not fishes; and certain fishes do not have all of them. The Mexican walking 'fish' (the axolotl, *Ambystoma mexicanum*) has gills and arms and legs, and is actually an amphibian, not a fish. Sharks and lampreys are fishes that lack a skeleton of bone, instead having a cartilage skeleton. Indeed, many primitive extinct jawless fishes do not have fins of any kind, and possess only the simplest rudimentary type of tail. It is true that all fishes have gills, and all have a notochord – a stiff fibrous rod of tissue that supports the backbone and often disappears after the formation of the bone or cartilage units. The notochord also exists in many primitive fossil ancestors of the first fish, and creatures having this feature are lumped as a group called 'chordates', placed in the Phylum Chordata. Chordates thus include all backboned animals ('vertebrates') as well as several primitive creatures that share certain 'advanced' anatomical features with fishes[1]. So, if we probe back into the murky depths of primeval seas, when do we see the first true fishes appear and what makes them recognisable as such?

The answer to this has always been shrouded with mystery and, rather than becoming clearer, the plot thickens each year with many new discoveries or radical reinterpretations of old data. Charles Doolittle Walcott, famous for his discoveries of the remarkable Burgess Shale animals of British Columbia, also recognised bone and scales from the Upper Ordovician Harding Sandstone in North America as far back as 1892. For many years these fossils, named as *Eryptychius* and *Astraspis*, went unchallenged as the oldest known vertebrate fossils, dating back to about 440 million years ago. Then in 1976, Frederik Bockelie and Richard Fortey announced the discovery of an Early Ordovician vertebrate, based on some bony fragments. These were named *Anatolepis* in honour of famous early fossil fish worker Anatol Heintz. Two years later John Repetski recognised even older fragments of *Anatolepis* in Late Cambrian strata, thus pushing the vertebrates, for the first time, back into the oldest geological period of the Phanerozoic Era, 540 million years ago to the present day. These authors all argued that the phosphatic exoskeleton of *Anatolepis* was similar to that of the younger armoured jawless fishes, the heterostracans.

Yet, it took only a few years after the publication of *Anatolepis* for other workers to denounce these finds as being non-vertebrates, more likely, they suggested, fragments of well-preserved arthropod appendages. The critics argued that the spiny ornament present on *Anatolepis* was well outside the known range of variation seen in fossil agnathan armours, and that the histological work was inconclusive. Their final (as they thought) nail-in-the-coffin was that the actual shape of the 'bony bits' of *Anatolepis* had no identifiable resemblance to that of early fish armour, and was more consistent with the spines of an arthropod.

Yet the idea of the presence of fish in the Cambrian was not buried by such persuasive doubters. The debate was thrown wide open again in the mid-1990s, amid a sea of new discoveries of even older, vaguely fish-like fossils from the Early Cambrian of China. What's more, the enigmatic group of worm-like creatures, the conodonts, were also thrust into the race for early vertebrate ancestry. Conodonts had long been known only from tiny phosphatic jaw-like fossils, but until the 1980s we had no idea what the whole organism looked like. We now know that they had worm-like bodies with tails supported by stiffened rods, and had V-shaped myomeres (muscle groups) along the body, and possessed a notochord, so in many respects they were fish-like creatures[1].

All this debate over the affinities of conodonts climaxed in a very confusing and emotional debate about exactly what constitutes a vertebrate. Does it lie in its tissue types or its gross morphology? Conodonts have a limited data set, mostly reliant on histological similarities with vertebrate dentine and a few gross characters seen in the rare whole specimens, that indicate vertebrate affinity. Yet the oldest chordate-like creatures, newly discovered from the 525-million-year-old Chengjiang site in Yunnan, China, include two well-preserved whole organisms. One of these, *Yunnanozoan*, another worm-like creature with supposed notochord and V-shaped muscle bands, was touted as being a possible Early Cambrian chordate (see Chapter 3).

Another, better preserved worm-like creature from Chengjiang was described about a year later by Shu and colleagues[2]. Named

Cathaymyrus, this beast looks, as we have shown in Chapter 3, to be a good Early Cambrian chordate, as it comes complete with what has been interpreted as a notochord and gut, pharyngeal gill slits and sigmoidally curved myotomes. At the same time it casts doubt on *Yunnanozoan* as having any relationship with the chordates. Despite the debate over the interpretations on these fossils, the only real, recognisable criterion for positively identifying an early creature as a fish, rather than a protochordate or fish-like ancestor, is the presence of true bone or cartilage. These are both derivatives from neural crest mesodermal tissue, and in recognisable bony form or shape are consistent with the known patterns of skeletal features seen in other early fishes. Mesodermal tissue is formed in the embryo and develops into bone and muscle types formed only in vertebrates. Thus, although microscopic fragments might fit the bill histologically, unless the actual shape of the bones match those of fish, then how do we know we are not dealing with a parallel lineage of creatures that evolved bone-like tissues without possessing the other anatomical refinements that befit a true, advanced vertebrate?

One characteristic feature of chordates seen in these enigmatic fossils from China and in conodonts is the presence of a series of V-shaped bands of muscle along the body, dividing the tail into segments. The tail in humans is secondarily lost in the evolution of apes from monkeys. These segments, called somites, have a close corresponding numerical relationship between the backbones (vertebrae) and somitic muscles. This is a prime feature of all chordates. Perhaps the most significant characteristic of most chordates is the ability to secrete phosphatic hard tissues, including that most advanced tissue of all, bone.

Aside from the fishes and higher vertebrates, the primitive groups of animals containing these characteristics include the urochordates (tunicates, or sea-squirts), cephalochordates (lancelets and related forms), and the puzzling conodonts, an extinct group of worm-like creatures known mainly from their phosphatic microfossils, but in recent years from some superb complete body fossils. In addition there is a mixed bag of bizarre early fossils simply termed 'problematica', some of which could have affinities to the vertebrates. All of these creatures lived in the sea, and this is undoubtedly the place where the first great evolutionary steps towards higher vertebrates took place. Now let us examine some of the main contenders in turn.

If you are down at the beach and you pick up a sea-squirt, it will do exactly as its name suggests – squirt out a jet of water. This ability to squirt is due to their well-developed muscles around a large central cavity, known as the pharynx. Sea-squirts feed by taking water in through the mouth and filtering it for food items by passing the water through their gill slits. Their other common name, 'tunicate' comes from the fact that they are embedded in a tough outer tunic of cellulose – the very same substance that gives plants their internal support. There is no denying the startling lack of similarity between an adult sea-squirt and a fish, yet it is during the sea-squirt's juvenile phase, when it exists as a larva, that it closely resembles the shape of a primitive fish. The larva of a tunicate has a long muscular tail, is supported by a notochord and has a spinal nerve cord. The head end has various sensory organs to enable the creature to swim and keep its bearing with respect to gravity and the direction of light. When this little tadpole-like creature finds a suitable place to settle, it head-butts itself on to the nearest rock by means of three hair-like sticky structures on the head (the papillae), and begins its metamorphosis into the blob-like adult form. The adult is immobile for the rest of its life, and as it develops it resorbs its long tail for nourishment. Tunicates can develop into either sex as they mature. They may even prefer to go AC/DC and reproduce asexually by budding off new animals. Pity we didn't retain this ability through to higher vertebrates as it would certainly have eased the burden of childbirth and put revitalised meaning into the old expression 'chip off the old block'!

Fossils thought to be of early tunicates are known from the dawn of the Palaeozoic Era, some 540 million years ago, although there is still much debate among scientists as to whether they really are tunicates or belong to completely new groups. *Palaeobotryllus*, from the Late Cambrian of Nevada, has a bubble-like form closely resembling modern colonies of the tunicate *Botryllus*. Microscopic platelets of some other strange creatures have been compared with the spicules found in the tunic of modern tunicates, and are also thought to represent ancient sea-squirts.

A problematic form described back in 1989 from the Early Ordovician of China by Clive Burrett, at the University of Tasmania, and myself has a phosphatic tubular exoskeleton with large blisters forming tubercles on the inside of the tube[3]. We named this form *Fenhsiangia* (after the town of Fenhsiang, in Hupei Province, China), and thought it must be somehow allied to the first vertebrates because vertebrate bone is the only tissue known to develop tubercles of this kind. However the tube-like shape of *Fenhsiangia* gives no clues as to the nature of the organism or its lifestyle. Because its tubercles are on the inside of the tubes, rather than on the presumed external surfaces, it remains a complete enigma, beyond meaningful comparison with all other known early fossil vertebrates. Yet such forms give us a tantalising glimpse into the myriad complexities involved in the development of primitive vertebrate tissues.

The lancelets (or cephalochordates) have been classically thought of as the closest ancestors of vertebrates because they are small, eel-like animals that have well-developed muscular somites with V-shaped muscle bands, a well-developed pharynx with numerous gill slits, and fin rays supporting a long median dorsal fin. Their long bodies are lance-shaped much like a primitive fish, hence their name. Only two living genera are known, *Branchiostoma* (once called *Amphioxus*) and *Epigonichthys*, growing to about 7 cm long. The sessile adults lie buried in the soft, sandy, shallow sea floor with their mouths protruding above the sand to take in food from the passing seawater. Lancelets have separate sexes and breed by shedding sperm and eggs into the water. Their fish-like larvae swim by motions of their powerful tail, yet they have fewer gill slits than their adult form. A circular ring of small tentacles, or cirri, surrounds the mouth, creating a current of water to enhance feeding. Lancelets lack a heart, but have a remarkable blood circulation system that is very close to that of typical vertebrates in possessing a large central artery, the ventral aorta. The fossil record of lancelets is poor, but nonetheless includes some well-preserved examples. *Palaeobranchiostoma* is known from the Permian of South Africa, and it closely resembles the living lancelets but has a larger, well-developed ventral or belly fin, and the dorsal fin is also larger and invested with numerous small barbs[1].

The next group, the conodonts, are more vertebrate-like than lancelets or tunicates because they have acquired the primitive phosphatic tissues that come close to being bone, or, as some have argued, actually are anatomically identical to some vertebrate tissues. Conodonts (from the Greek, meaning 'cone teeth') are known principally from tiny microscopic remains of phosphatic jaw-like structures, dating back into the Late Cambrian, about 520 million years ago. Thousands of such minuscule fossil remains can be picked out from Palaeozoic and Triassic limestones by dissolving the rock in weak acetic acid and sieving off the undissolved phosphatic elements, mostly bone fragments and conodonts. The shape and form of conodonts have been intensely studied from their isolated remains and used in correlating and assessing the ages of rock sequences. However, until quite recently, we had no inkling what the creatures that possessed them looked like. We call these tooth-like remains of conodonts, the conodont 'elements', and refer to the organism as a whole creature as the conodont 'animal'.

The conodont elements can be simple rod or cone-like forms, blades with teeth-like protuberances, or complex platform shapes. Sets of these elements often occur together, and thus each conodont animal possessed an assemblage of differently shaped conodont elements. Studies under the microscope of the conodont tissue types have hinted at vertebrate affinities, and recently a paper published by Ivan Sansom and his colleagues from London in the American journal *Science* suggested that true bone cells were present in some conodonts[4]. However, not all palaeontologists agree with this hypothesis. The jaw-like appearance of some conodont elements is deceptive, as tooth-like cusps along the ridges at first sight never appear to show any sign of wear, as teeth would normally do. This means that the structures may have merely been supporting gills or filter-feeding devices, and not used for food reduction in any direct way. However, Mark Purnell of the University of Leicester, England, has identified microwear patterns on conodont elements, suggesting that food was crushed and sheared between opposing elements in the same way that vertebrates use teeth. Furthermore, he extended the hypothesis to argue that conodonts, as the first vertebrates, were predators[5]. Comparisons with vertebrate dentine, a tissue found in the teeth of higher animals, but once widespread over the entire external skeleton of early fishes, make the point that such tissues were widespread within the lower vertebrates and lower chordates. Yet, this doesn't necessarily prove that conodonts had true teeth.

In the early 1980s the first fossil remains of whole conodont animals were found in the Granton Shrimp Beds near Edinburgh, of Early Carboniferous age, about 340 million years old[6]. These show that conodonts were long, worm-like creatures with tails having supporting fin rays. In the head region they were found to possess a cluster of little conodont elements. This evidence, together with new data on the 'bone' structure of conodonts, is powerful evidence that they were close to the evolutionary line leading to true vertebrates. If the possession of true bone is sufficient grounds to call a creature a vertebrate, then conodonts could indeed be classified as early fish.

Perhaps more convincing is the probable presence of cartilaginous sclerotic rings or eye capsules in conodonts reported from the Ordovician Soom Shale of South Africa by Sarah Gabbott and Dick Aldridge, of Leicester University, and Johannes Theron, of the South

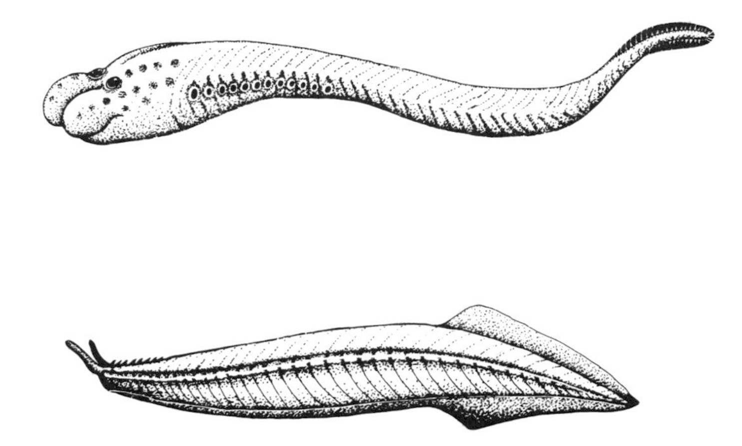

Show us your backbone! These two creatures, each about as long as a cigarette, represent two early experimental lines that could have led towards the first back-boned animals, the vertebrates. The top one, known affectionately by us as 'the amazing bum-faced wonder', is a reconstruction of a conodont animal (after Dick Aldridge, *his work, not the face!). The lower one is* Pikaia, *a possible protochordate from the* Cambridge Burgess Shale *fauna (after the work of* Simon Conway Morris*).*

African Geological Survey, in 1995, and also identified in the few other conodont animal fossils known[7]. The eyes of early fishes invariably lacked scleral coverage, except in the most advanced of the jawless fishes, the highly developed osteostracans and even more advanced jawed fishes. The beautifully preserved conodont animal called *Promissum* from South Africa was enormous, nearly 40 cm long, much larger than previous whole body fossils from Edinburgh[6]. They even had muscle tissue preserved. Thus, as Philippe Janvier, of the Museum National d'Histoire Naturelle, Paris, has pointed out, conodonts may not only contend to be fishes, but they could even claim to be more advanced fishes than most of the bone-covered, jawless forms so far described from the Early Palaeozoic[8].

The first true fishes were creatures capable of powerful locomotion, because the outer covering is invested with bone upon which muscles can attach internally. No matter what was going on at the business end of the fish, the simplest defining character of all fish – the presence of bony plates and scales covering the outer skin – must really be taken as the benchmark feature separating fishes from other chordates. However, as they lack these characteristic, and somehow aesthetically beautiful, sculptured bony plates, I prefer to place conodonts one peg lower than fish on the evolutionary scale.

There is one other contender for the mantle of the almighty ancestor of all time, the mitrates. These are an extinct group of echinoderms: the phylum containing starfishes, sea-urchins and sea-cucumbers amongst others. The view that early fossil mitrates were closely related to the first vertebrates, championed by Dick Jefferies of the Natural History Museum, London, has not received much support in recent years. Most workers in the field today regard mitrates as simply an interesting group of echinoderms. They lack bone, a notochord and gills; and resemblances to early vertebrates proposed by Jefferies, such as the muscular tail, are regarded as features that evolved within the echinoderms in parallel to vertebrate evolution, rather than as a step towards the lineage leading to true fishes.

Another proposed candidate for the oldest cephalochordate, and thus one of the most ancient ancestors of all vertebrates, could be *Pikaia*, from the Middle Cambrian Burgess Shale of British Columbia, in Canada. Although superficially worm-like in appearance, *Pikaia* has a number of features that point to its chordate affinities. It has what appears to be a notochord, a rod of stiffened fibrous tissue supporting the axis of the animals. *Pikaia* has a body form and overall anatomy similar to modern lancelets; however, it has yet to be fully studied in detail so its evolutionary position remains uncertain. The recent discovery of an even older cephalochordate from the Early Cambrian of China, *Cathaymyrus*, shows that *Pikaia*-like creatures were around from at least 530 million years ago[2].

So, from this look at living, primitive chordate-like creatures, together with the scant but interesting fossil record of these groups, the most convincing resemblances to the first fishes are seen in the larval stages of several invertebrate groups. By the process of paedomorphosis, that is the retention of juvenile features into a descendant adult phase, as the juvenile sexually matures earlier, the larvae of either tunicates or cephalochordates could have quite readily developed into a primitive 'boneless' fish. This provides a mechanism for the evolution of the first fishes. It may well have been little more than a quirk of development, such as a slight hormonal hiccup in the timing of the onset of maturity, rather than necessarily reflecting the initiation of any major environmental changes that favoured the first protochordates over all the other bizarre and wonderful life forms abounding at the same time. As Steven Jay Gould has suggested in the epilogue to his book *Wonderful Life*, the lucky rise of the first vertebrates may well be due to the fact that creatures like *Pikaia* (or *Cathaymyrus*) by some quirk of fate simply survived whatever events or factors caused the decimation of the Burgess Shale faunas.

The identification of true vertebrates in the early fossil record thus relies heavily on the definition of what exactly bone is. Fishes have bone in which there are cell-spaces for bone-producing cells. The external layers of the most primitive of all fish bones have an ornament covered by a thin enameloid layer over a dentine layer. The layer of non-cellular bone, found in some fossil jawless fishes is called 'aspidin'. Thus our humble human teeth are our only remaining vestige of that first outer covering of complete tooth-like squamation. One day it invaded the mouth cavities of jawless fishes, and presumably some took root to become real teeth. But that is another story (Chapter 6).

Bone is very much the key to understanding the success of the first fishes. It provided a solid support for the attachment of muscles, giving a greater efficiency for using a muscular tail to propel the creature through water. Escape from predators and the improved ability to catch slower-moving prey must have been huge adaptational advances over previously slow-moving protovertebrate-like creatures. We explore the importance of predation pressure in driving evolution in Chapter 16. Bone not only increases speed but also acts as a storehouse for phosphates and other chemicals required in daily metabolism. But there's more! It gives protection to the most vulnerable parts of the animal's anatomy, such as the brain and heart, enabling the organism to have an even greater chance of survival after an encounter with a predator. Once bone evolved in its refined state, fishes underwent a rapid evolutionary explosion of diversity.

So, back to the original question, that came to mind as I stood on the stormy summit of Mt Watt back in August 1993. Where did the oldest known fishes originate? The oldest identifiable fossil vertebrate, one that has undoubted real bone, comes from the Early Ordovician rocks of central Australia, dated at about 485 million years old, from both Mt Watt and Mt Charlotte, near Alice Springs. These are the impressions of simple fishes called *Arandaspis* and *Porophoraspis*, described by Alex Ritchie and Joyce Gilbert-Tomlinson in 1977[9]. These point to Gondwana (the great southern continent comprising the now separated continents of South America, Africa, Antarctica, Madagascar, India and Australia) as having been the most likely place for the origin of all vertebrates. The next oldest, but much better preserved fossil fishes, also known from relatively complete remains, come from the Late Ordovician of Bolivia, in South America, also part of the

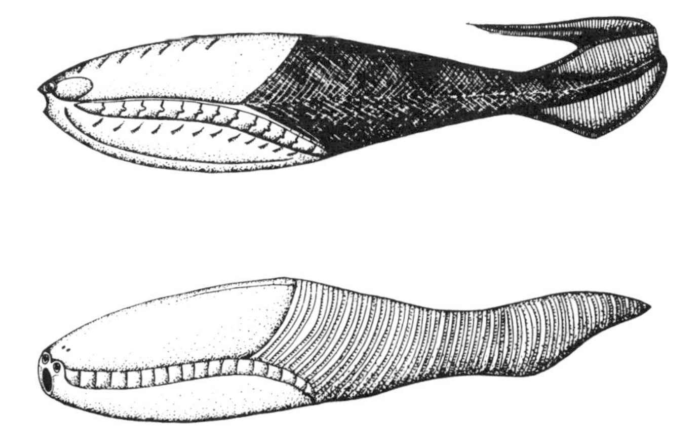

The ones that didn't get away. The world's two oldest known relatively complete fishes from the Middle–Late Ordovician. Top, Sacabambaspis, *from Bolivia and Australia (after Pierre Yves-Gagnier); below* Arandaspis, *from Australia (after Alex Ritchie). Both fish were about as long as your foot, and had complete outer coverings of dermal bone.*

ancient supercontinent of Gondwana[10]. These fishes, described as *Sacabambaspis*, are very similar in form to *Arandaspis*. Both the Australian and South American forms have a large dorsal bony shield and a large ventral shield separated by numerous rectangular plates which covered the 12–15 pairs of gill openings, and headlight-like orbits facing to the front of the animal. The tails were simple, covered with long comb-like trunk scales, with a very simple tail fin, equally developed both above and below the main axis of the body in the one form it is known. Quite recently *Sacabambaspis* body scales have been identified from the Ordovician of central Australia by Gavin Young[11].

However, as I write this, the presses keep rolling off news flashes that shatter these accepted foundations of early vertebrate orgins. In late October 1996, Gavin Young of the Australian Geological Survey Organisation in Canberra, and his colleagues, published a report in *Nature* of a possible Late Cambrian vertebrate from central Queensland, Australia, dated at about 520 million years old[12]. The spectre of *Anatolepis* rises once more. These bony fragments differ from those of *Anatolepis* in showing that they possessed an extensive pore-canal sensory system, a feature found in all true fishes. The histology of these bony pieces shows a three-layered skeleton which has a thick enameloid outer layer but lacks dentinous tissue, once thought to be a definable feature of all early vertebrate tissues.

Furthermore, where once the origins of large bony plates in fishes were thought to have formed from small centres (called odontodes) which joined up to make large bony areas, the new material shows no such mechanism of formation. On the contrary it supports an opposing interpretation, that bony sheets could well have been a primitive pattern for fishes, and that enameloid tissues preceded the widespread evolution of dentine support tissues in the dermal skeleton. The new fossils from central Queensland also show diversity – two kinds of different phosphatic dermal armour, suggesting experimentation of bone formation in these first fishes or fish-like animals. In concluding, Gavin Young and his colleagues say that a pore-canal system with enamel-like hard tissues may well have preceded the formation of cellular bone with dentinous tissues in vertebrates. The implications of this are that conodonts and *Anatolepis* could well represent divergent lineages within the first radiations of vertebrate hard tissues, and so may not be so close to fishes as some have passionately argued.

Perhaps the most startling revelation from the fossil scales that Gavin Young has described from central Australia is the possibility that not all of the microscopic bony remains might come from primitive jawless fishes. One scale type named *Areyongia* closely resembles the previously oldest known shark scales which come from the Early Silurian of Mongolia. Young suggests that they were possibly a primitive precursor to the chondrichthyan (sharks and their allies) *Polymerolepis* type of scale, yet as their internal structures show significant differences he was not confident of placing them within the Chondrichthyes – to do so would not only dramatically push back the origins of sharks by some 30 million years at least, but it would rattle the very foundations of the origins of toothed vertebrates.

The origins of the first fishes appear to be firmly rooted within the ancient Australian continent, and supporting evidence shows that material from South America, slightly younger than the central Australian fossils, places Gondwana, the great southern supercontinent, as the birthplace of the first fishes.

5

Walk on the Wild Side

The colonisation of land

It had been a long drive. Here I was, eight hours after leaving Perth, and driving through low scrubby vegetation on a yellow sand track. Winding between stands of banksias and grevilleas in Kalbarri National Park, I was close to our destination – the spectacular Murchison River gorge. Here the river had cut through hundreds of metres of blood-red sandstone, apparently with the ease of a knife cutting through butter. The reason I was here was because in peeling back the layers of time, the river had revealed tantalising indications of what would turn out to be evidence of some of the first animals that ever walked on land.

It all started in 1979, a few months after I joined the Western Australian Museum. One spring morning a letter landed on my desk from the then Director of the Western Australian National Parks, Francis Smith. He was about to retire, and clearing his desk he found a photo that he had taken some years earlier of a site at the bottom of the Murchison River gorge which had long perplexed him. I took one look at the photo and realised that the view of a large, flat, ripple-marked surface covered by what appeared to be motor-cycle tracks was something very special. Either Harley Davidsons had been around in the Late Silurian, more than 400 million years ago, or some pretty huge arthropods had once been wandering around the landscape of ancient Western Australia.

Accompanied by the park ranger, Cec Cockman, I set off early the next morning from our camp site in a sand pit not far from the gorge. It was promising to be one of those ferociously hot October days, with shade temperatures in the 40s (centigrade). The trouble was there wouldn't be too much shade where we were going. The track we were driving along got narrower and narrower as we got closer to the gorge.

'Stop here', shouted Cec over the sound of the straining vehicle engine. 'We're nearly there – just down that goat track to the bottom of the gorge, then a few hundred metres along and we'll be at the site'.

Goat track? Nice one, Cec. Even goats would need crampons to get down there safely. But we outdid the goats and eventually reached the bottom, carried down as much by the expectations of what we might find as by the pull of gravity. Its hard to describe my first feelings on what we saw, at this amazing site. What can you feel when you are confronted by the results of a few minutes activity made by long-extinct animals more than 400 million years ago? Humble? Amazed? Excited? Speechless? Probably all these feelings and more ran through me as I looked at an 8 × 5 metre flat pavement of ripple-marked sandstone across which ran half a dozen trackways nearly 20 cm across, each a good 5 metres long. I had neither seen nor heard of anything comparable anywhere else in the world. But I had seen a very similar looking track, less than a metre long, far away in Norway some years earlier. They had been made by a large scorpion-like animal called a eurypterid. Could these have been made by the same type of animal?

That visit was the first of many that I was to make to the area. In more recent years I have worked in collaboration with Nigel Trewin from the University of Aberdeen in Scotland. He had taught me palaeontology there in the late 1960s. That we would be working together on the other side of the world 25 years later was unexpected, to say the least; but decidedly fruitful. Nigel is an expert sedimentologist and trace fossil man with a great knack of being able to reconstruct ancient environments from looking at layers of sediment. Over the years not only have we found a veritable trace fossil menagerie, but Nigel has come up with an environmental interpretation that allows us to figure out how such tracks were preserved. More importantly, we have been able to demonstrate that these animals were among the first to take a walk on the wild side, and come out of the water on to dry land.

The sandstone in which this evidence is fossilised is known as the Tumblagooda Sandstone. It was laid down by a huge river system that drained inland Western Australia about 420 million years ago and discharged into a shallow sea. The world's geography at that time was quite different to what it is today. Then, the world's southern continents nestled together as one gigantic supercontinent called Gondwana. Southern Australia snuggled up against Antarctica, while a fledgling India was moored off Western Australia's southwest coast. Off the northwest coast of Australia bits of present-day southeast Asia, like Thailand, were just beginning to break free from the clutches of Gondwana and start their northward flight to freedom.

The 420 million-year-old pre-Murchison River would have drained a towering, naked mountain chain made of ancient Precambrian rocks. At this time plants were only just making their first, tentative attempts to take root on the land. Vascular plants (so-called higher plants, like daisies and pine trees, with water and food conducting tissues) had not evolved by this time. All that clung to windswept rocks and nestled in hollows were liverworts and mosses. Without an effective plant cover there would have been little to temper the powerful erosive forces of the wind, rain and sun. Millions of tonnes of sand would have been constantly torn from the mountains and swept away by raging torrents down into the sea. The air would have been filled with swirling, fine dust brought by sandstorms sweeping across the bare land. Ironically, clearing and overgrazing by man of this same, heavily eroded area today sometimes produces a similar effect – massive erosion and choking dust storms. It would have been like this every day on Earth more than 400 million years ago. Yet, rather surprisingly really, it was in just such a hostile environment as this that animals left their first calling cards on land – footprints on the sands of time.

The Tumblagooda Sandstone is one of those furiously frustrating deposits for a palaeontologist. Search as hard as you might, the fossilised remains of an animal's body never (or hardly ever) turn up. There has been one stunning exception, though, as I will reveal in Chapter 7. No, not a bone, nor a shell, nor the actual remains of a body. Occasionally you might find a tantalising suggestion of a place where an animal may have rested in some wet sand. But often it is more the eye of faith, or hope, than any realistic fossil.

But fossils do occur in these harsh sandstones, not as body fossils, but as somewhat more enigmatic tracks and burrows. Known as 'trace fossils', in many ways these tell-tale marks on the rocks that record the behaviour and activity of animals long, long ago, provide a more intimate view of life than the dead carcass of the animal itself. I have to admit that to me, as a palaeontologist, there are few things more fascinating than finding a slab of rock that is covered by rows of delicate imprints made by the patter of tiny feet of an animal as it scurried across

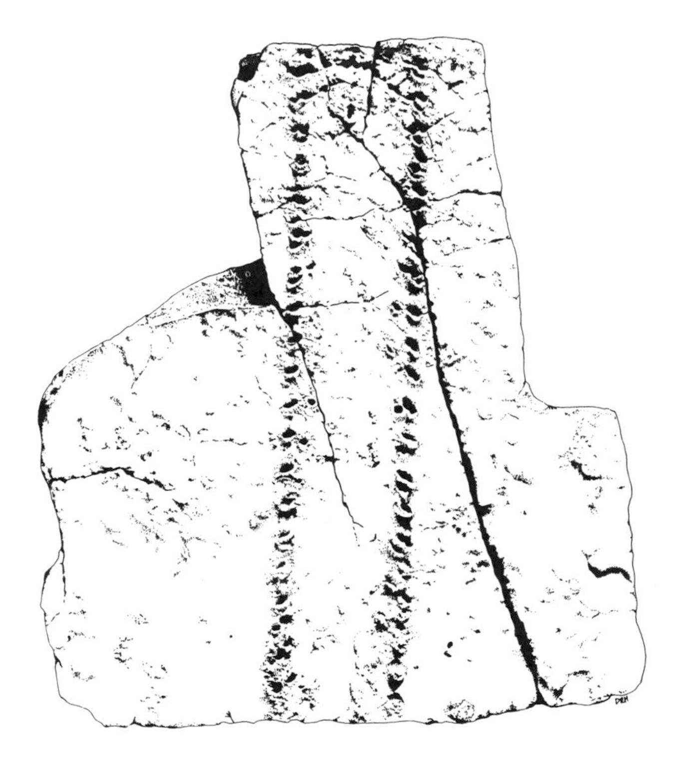

Footprints on the sands of time: possibly the oldest evidence of life walking on land. These are 420-million-year-old eurypterid trackways preserved in the Tumblagooda Sandstone from near Kalbarri, Western Australia. A small hand could span the track.

a damp sand flat hundreds of millions of years ago – a fossil snapshot in time. When I make such a find what I may be looking at is the activity of a single animal, for perhaps just two minutes of its existence, an unimaginably vast period of time ago. As the Reverend William Buckland* so eloquently wrote about fossilised trackways in 1836: 'Yet we behold them, stamped upon the rock, distinct as the track of the passing animal upon the recent snow; as if to show that thousands of

---

* William Buckland (1784–1856), Dean of Westminster, Professor of Mineralogy and Reader of Geology at Oxford was well renowned for his eccentricities. Buckland boasted that he had eaten his way through much of the animal kingdom, and that the worst thing was a mole, although he had also been known to argue that a blue-bottle fly came a close second. He is also reported (in E.T. Cook's *Life of Ruskin*) to have swallowed part of King Louis XIV's heart at a dinner party. Shown the heart, which looked remarkably like pumice stone, Buckland is alleged to have exclaimed that while he had eaten many strange things he had never eaten the heart of a king, and promptly swallowed the precious relic.

years are but as nothing amidst Eternity – and, as it were, in mockery of the fleeting perishable course of the mightiest potentates among mankind.'

The trace fossils that Nigel Trewin and I have been eagerly searching for in the Murchison River gorge over the last decade come in a bewildering array of shapes and sizes. Some can easily be interpreted as either trackways or as burrows. However, others are less easily explained, making it difficult to suggest what made them. Easiest to recognise are the tracks made by multi-legged arthropods (the group of segmented invertebrates that today includes animals such as insects, spiders and crabs)[1]. The tracks consist of parallel rows of spaced pits. These may cluster together in groups of three, or six, or sometimes eleven, and much of the research in the last few years has centred on trying, with the zeal of a modern Sherlock Holmes, to work out which animal made which of the footprints. Some, we think, were made by eurypterids. Popularly known as 'sea scorpions', these arthropods that grew up to more than 2 metres long, had six pairs of appendages (making them chelicerates, like today's king crab, *Limulus*). The back three pairs were used for walking or swimming, while the front three were often modified in some way to help in feeding. In some they occurred as scorpion-like pincers; in others they were huge cage-like structures that were probably used to trap their prey. Eurypterids were the top-line predators of the rivers and streams, as well as the oceans, 400-odd million years ago.

Other tracks were made by animals with more than three sets of walking legs, perhaps by huge centipede-like animals more than a metre long. One enormous set of footprints, more than 30 cm across, may well have been the only evidence we have in Gondwana of the existence of huge true scorpions that may have reached lengths of up to 1 metre. As I shall discuss shortly, the evidence all points to the Tumblagooda tracks having been left by animals walking out of water. And as such they represent the earliest direct evidence for animals walking on land. Yet, once again, it is surprising that we are dealing with such huge animals so early in the evolutionary history of some groups, in this case terrestrial arthropods. However, not all early arthropods were large. Much depends on where the animals lived and how they were preserved, as I shall elaborate more fully below.

One of the more important discoveries to come out of the work on the arthropod trackways is the realisation that some of these animals, thought to have been exclusively aquatic, were at least amphibious.

Gimme a kiss! Gruesome looking eurypterids like this one called Mixopterus, *nearly a metre long, were amongst the earliest animals to leave the sea and make short forays onto land.*

Some may even have been adapted to a fully terrestrial existence. Bear in mind, as I shall return to this point later on, that these animals were strolling along the shores *before* vascular plants had taken root on land. The oldest terrestrial rocks in which such land plants, like *Cooksonia*, have been detected are a little over 400 million years old. We are dealing with an ancient, windswept world in which there was a great diversity of large to small arthropods, that were scuttling, crawling and swaggering across sand flats and up sand dunes, long before the first animals with backbones had the necessary equipment to venture out of the water and up on to the beach. To substantiate this conclusion Nigel and I (mainly Nigel) had to look not only at the tracks themselves, but also the nature of the sandstone in which they were fossilised. This allows us to bring back to life the environment in which these animals were walking, breeding, eating, being eaten and dying.

The rocks that form the walls of the Murchison River gorge may appear at first glance to be a monotonous sequence of layers of sandstone, but with closer scrutiny it is possible to see that there is sandstone and there is sandstone. Some layers are thick and massive. In these we have found very little in the way of trace fossils. These sandstones represent the debris dumped by a fast-flowing river. Other, thinner layers between these river sands are often ripple-marked and represent vast sand flats laid down as the flooding river subsided. Other sandstones, which on closer inspection can be seen to be made of finer grains, are the remains of windblown sand – fossil sand dunes. And it is here, at the junction between these sand flats and windblown sands, that the trackways are most often preserved.

What the footprints reveal is that a range of arthropods may have been wandering between pools of water, perhaps as they evaporated in the harsh sun. Treading with their multi-legged feet over the wet sand, their imprints were preserved for over 400 million years by sandstorms that dropped their fine sand on the wet surface. Exposed hundreds of millions of years later by the action of the modern Murchison River carving through the ancient layers of sandstone, the rocks preferentially split between the waterlain and windblown sands, revealing the fossil trackways. At one spectacular site in the gorge, located by Roger Hocking of the Geological Survey of Western Australia, a steep slope from a sand flat into what was once a pool of water has been preserved. Trackways can be seen to wind their way along the sand flat to the top of the slope. Then they are replaced by a pair of deep grooves made by the arthropod, as it slid, crocodile fashion, down the bank and the pool.

But the clinching argument for such tracks having been made out of water came from one of the most unusual trackways that Nigel Trewin and I found. Late one afternoon in 1995 we were walking along the side of a small gully that ran down to the main Murchison River gorge, one that we hadn't explored before. Just as we were about to give this gully up as a bad bet, Nigel suddenly spotted a good arthropod trackway on a ledge overhanging the side of the gully: parallel rows of clusters of delicate imprints and a central drag-mark – made by some sort of unknown arthropod. But there, right next to it, was another, the like of which we had never seen before in 16 years of searching. Equally fascinating was that at the end of each footprint was either a little pile or trail of sand, made as the animal lifted its foot out of the cloying, wet sand. As the feet rose up, gravity took hold and a mixture of sloppy wet sand and water fell down to leave a miniature sandcastle. The fact that such a moment in an animal's activity made so far back in deep time could be preserved is really quite miraculous. Such preservation could only happen out of water – in air, as the Sun blazed down and as the wind spread a gentle cloak of dust over the little sandcastles, hiding them from the light of day and preserving them almost for eternity. But hundreds of millions of years later the wind, water and Sun conspired once again, this time to release them from their almost timeless cover. What was it that Longfellow wrote in his 'Psalm of Life'?:

> We can make our lives sublime,
> And, departing, leave behind us
> Footprints on the sands of time.

There can be little argument that the first steps taken by animals on to land hundreds of millions of years ago was one of the most significant events in the history of life on Earth. Pinpointing the time exactly when these first faltering steps on land took place, and discovering what type of animal made them are obviously very difficult. The trace fossils point to at least 420 million years ago. The Tumblagooda tracks are not unique. Recently, small tracks of a similar or even slightly older age, possibly made by a millipede-like animal, have been described from the Borrowdale Volcanic Group in the Lake District in England[2]. Like the Tumblagooda tracks, these have been interpreted as having been made by arthropods walking out of water. And like the Tumblagooda sequence, the associated sediments are freshwater.

Both occurrences lend support to the notion that colonisation of the land by animals was not from the sea on to land, but from rivers and streams. But what other evidence is known, more indirect, or direct in the form of body fossils? During the last decade a number of significant new fossil discoveries have been made in addition to the Tumblagooda tracks, that have revolutionised our ideas on the manner in which terrestrial ecosystems became established.

Ingrained in our consciousness from an early age is the idea that the sequence of colonisation of the land from the sea and rivers saw its parallel in the evolutionary tree of life – firstly plants colonised the land; followed by small animals that grazed on these plants; and then these in turn were followed by larger, carnivorous animals that fed on the herbivores. But when we look at the fossil record, there is little support for such a scenario. Recent fossil finds made from a range of sites, from Shropshire in England to Sweden to the USA, impart a story that is more complex and in many ways more fascinating.

Establishing which animals first set foot on land is naturally heavily constrained by the character of the fossil record. While an aqueous environment is likely to be often conducive to fossilisation, the bare, hostile land surface on which these early animals chose to walk was far from an ideal place in which to become fossilised. The land would have been whipped by winds; water-courses would have constantly shifted; while sandstorms repeatedly covered, uncovered and moved dead bodies. However, at a few places, a combination of serendipitous factors came together to produce exceptional preservation. One such example was in the north of Scotland at Rhynie, where in a fossilised hot spring environment about 395 million years old, the oldest known well-preserved terrestrial ecosystem occurs. Here the detailed interrelationships between animals (all arthropods), plants, cyanobacteria and fungi can be interpreted.

One important question to be answered is whether the land surface on which the animals first walked was carpeted by plants. If so, just what were these plants? Or was the land upon which these animals traipsed bare? What we know about the earliest land plants is shrouded in almost as much mist as our knowledge of the first animals. The earliest evidence is a collection of spores found by Jane Gray of the University of Oregon in Ordovician rocks up to 470 million years old from Libya[3]. These spores resemble those of living mosses, liverworts and ferns, suggesting to Gray that some of the earliest land plants were liverwort-like. As such they would have been able to cope with long periods of drought. But even older evidence of life on land has been uncovered in recent years, in the form of bacterial and algal mats preserved in 1200 million-year-old cherts in Arizona.

The earliest indications, albeit rather indirect, of animal activity on land have been interpreted from Late Ordovician (450 million years old) fossil soils from Pennsylvania. Within these ancient soils Greg Retallack and Carolyn Feakes of the University of Oregon discovered deep, vertical burrows, up to 20 mm in diameter. They interpreted these as possibly having been made by millipedes[4]. More indirect evidence for animals on land has come from 410 million-year-old Late Silurian rocks in Sweden. Martha Sherwood-Pike, also of the University of Oregon, and Jane Gray found what they interpreted as fossilised faecal pellets. These contain fungal hyphae, implying the existence of a fungivorous microarthropod, perhaps a mite or a millipede[5]. If correct, this suggests that some of the earliest land animals occupied a decomposer niche, and thus played an important role in the establishment of soils. Ecosystems like this would have played a crucial role in the formation of soils, reworking the upper layers and increasing nitrate and phosphate levels. This would have provided a necessary habitat to allow subsequent colonisation by vascular plants.

The earliest vascular plants were simple forms like lycophytes such as *Baragwanathia* and the rhyniophytes *Salopella* and *Hedeia* from Victoria in Australia, lycophytes and *Psilophyton* from Libya, and *Cooksonia* – perhaps the most primitive-looking of all plants. These all appeared in the fossil record in sediments formed in Late Silurian times, around 400 million years ago. Doubts had been raised about whether *Cooksonia* really grew on land in the Silurian. *Cooksonia* had long been thought to be the earliest land plant. Confirmation of its terrestrial status was provided by Dianne Edwards and her co-workers from the University of Wales at Cardiff, with the discovery of water-conducting vessels (tracheids) and stomata in *Cooksonia* from Late Silurian rocks in the Ludlow Bone Bed in Shropshire. Only in land plants do these structures occur.

Baragwanathia and *Cooksonia* are usually preserved in marine sediments, indicating that they grew nearshore. To assess the degree to which inland sites were vegetated at this time we must turn to the classic Early Devonian site at Rhynie in Aberdeenshire, Scotland. About 395 million years ago what is now a small field in Aberdeenshire looked more like New Zealand, with hot springs, belching fumaroles and geysers. In this environment a range of plants and animals were trapped in small pools, giving us an intimate view of life on land so long ago. In addition to early vascular plants, there were cyanobacteria, algae and fungi. Even mycorrhizal fungi (fungi that grow in assocation with roots of plants) have been identified, making this the earliest evidence of an association between fungi and plants. The oldest known terrestrial lichens have also been found at Rhynie. And crawling through this undergrowth was a rich fauna dominated by arthropods: the fairy shrimp *Lepidocaris*; three species of spider-like trigonotarbids; the mite *Protocarus*; the springtail *Rhyniella*. The presence of book lungs in one trigonotarbid, *Palaeocharinus*, provides firm evidence that the animal was truly terrestrial[6]. Book lungs are internal, layered structures used by living scorpions and some spiders for respiration.

Early animal land pioneers would obviously have faced immense problems in making the transition from an aqueous to a dry environment. The land was a hostile place to animals that had evolved in water. Profound physiological changes were needed to allow a transition to a dry, terrestrial habitat, where daily temperature changes were much greater than in aquatic environments, and respiration was carried out in the air rather than under water. One of the major problems faced by these animals was to avoid drying out. The indication of millipedes very early in terrestrial colonisation suggests that they may have initially coped with problems of temperature and desiccation by burrowing in the 'soil', or hiding in any cracks or crannies they could find, much as they do today.

To survive such a hostile world, animals would have needed to have been preadapted to terrestrial conditions in some manner. All the evidence to date, from both trace and body fossils, points to arthropods as being the first group of animals to colonise the land. One of the main reasons for their success lay in their tough, crusty outer shell, which evolved at the beginning of the Cambrian Period some 540 million years ago. This preadapted arthropods to a life on land. As well as conferring defence, it gave strength to the body for locomotion and feeding. It also helped overcome the dragging effects of gravity, a big problem to any animal used to being buoyed up by water. Without such pre-strengthened legs, walking around on land would have been very difficult. This is shown by the fact that land arthropods can still be vulnerable to mechanical failure after moulting while their new cuticle is hardening. Another anti-desiccation factor possessed by arthropods was their ability, like land plants, to secrete a waterproof outer covering on their shell to stop water loss.

Perhaps the most exciting discovery to come out of a range of new fossil finds in recent years is the realisation that early terrestrial arthropods were predominantly carnivores. During the processing of a muddy siltstone from just above the famous Ludlow Bone Bed in Shropshire, Andrew Jeram, while at the University of Manchester, uncovered a large quantity of arthropod cuticle[7]. Masses of bits of legs, bodies and even a few articulated specimens turned out to represent at least two types of centipedes, a trigonotarbid arachnid (described recently by Jason Dunlop of the University of Manchester as *Eotarbus jerami*), an arthropleurid and a probable terrestrial scorpion. Dated as late Silurian (414 million years old) in age, these are the oldest body fossils of land animals. *Eotarbus* is the earliest known non-scorpion arachnid[8].

Slightly younger, and superbly preserved remains some 375 million years old, have been found in recent years at Gilboa in New York State. Here, a similar arthropod-dominated terrestrial fauna of millipedes, mites, pseudoscorpions and arachnids occurs, including the earliest

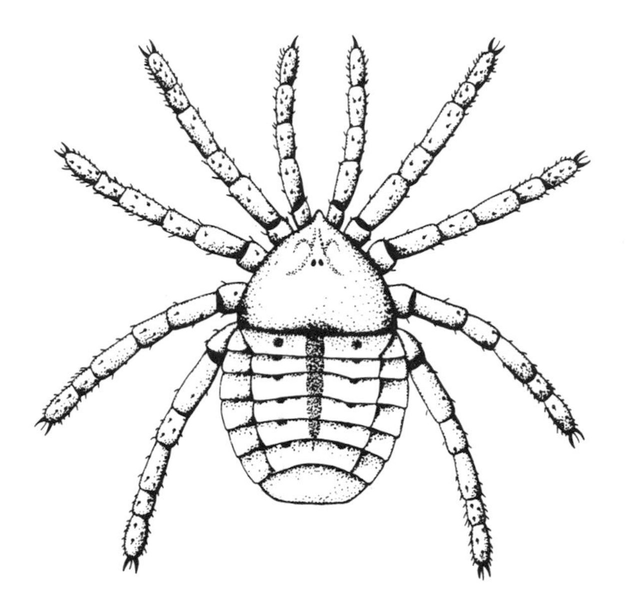

Well before Miss Muffet *had evolved, simple creatures like this early arachnid, a trigonotarbid called* Eotarbus jerami, *were crawling on the land about* 400 *million years ago.* Not *a giant of the past, the beast would have had trouble peering over a grain of rice.* This *drawing is based on an original by* Jason Dunlop.

known spider[9]. Like modern spiders this tiny animal, between half and about one centimetre long, called *Attercopus fimbriunguis*, is equipped with a spinneret, implying that spiders as old as this one had the ability to spin webs. It was also endowed with fearsome fangs and poison gland[10]. The name *Attercopus* comes from an Old English word for spider – attercop (literally, poison-head). The pseudoscorpion also represents the earliest known representative of this group of arachnids. Named *Dracochela*, this form is surprisingly similar to living pseudoscorpions, sufficient for living pseudoscorpion expert Mark Harvey at the Western Australian Museum to place it within a living superfamily.

As at Rhynie and Ludlow, the Gilboa arthropods were mainly predators, with a few detritivores. No herbivores are known. The predators probably fed mainly on the microarthropod detritivores such as mites and millipedes. These are rarely preserved, but probably made up a substantial part of the litter fauna. Surprisingly, there is little evidence for herbivorous arthropods in the fossil record until well into the Carboniferous Period, about 330 million years ago. William Shear of Hampden-Sydney College in Virginia has suggested that by-products

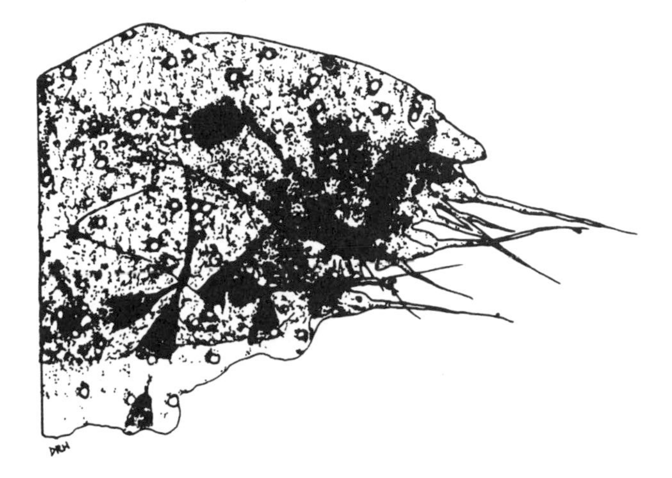

The amazingly beautiful and somewhat sensual backside of an early fossil spider from the Late Devonian *of* New York State, *showing off its spinnerets.* (*Greatly magnified sketch*).

from the synthesis of lignin, which occurred in early vascular plants, may have been toxic to early terrestrial animals. He argues that perhaps true herbivory only evolved when animals had evolved enzymes and a gut microflora of symbiotic bacteria, and were capable of breaking up fresh plant material directly without the need for external decomposers[11].

Persuasive as this argument might be, new discoveries by Dianne Edwards, Paul Selden and colleagues have cast some doubt on this interpretation. They have identified coprolites (fossil faeces) from Ludlow and from slightly younger rocks also in the Welsh Borderlands, that imply animals were feeding on plant matter at this time. These ancient droppings consist largely of undigested land-plant spores. This suggests either that an unknown herbivorous animal was around, quietly chewing on early land plants, or that detritivores (such as millipedes) could have been responsible. The question is, therefore, were the animals feeding on living or dead plants? If they were feeding on living plant material it fits in with some ideas that pollen feeders preceded leaf feeders in arthropod evolution, as spores and pollen have much higher nutrient levels. It is perhaps most likely that detritivores were feeding on litter rich in spores. Certainly, there is little evidence for animals feeding on leaf matter during the Devonian and Carboniferous.

There are, surprisingly perhaps, some close parallels in the pattern of ecological succession seen today compared with what might have been occurring as arthropods colonised the land hundreds of millions of years ago. The great success of predatory arthropods in colonising barren landscapes has been documented in ecological succession at places like Mount St Helens and Krakatau, in the aftermath of cataclysmic volcanic explosions. Zoological expeditions to the Krakatau Islands in 1984 and 1985, led by Ian Thornton of La Trobe University in Victoria, showed convincingly that arthropods rapidly colonised areas of recently formed, bare, ash-covered lava. Significantly, they did so well before the establishment of any plant. Not only that, but the first colonisers were much the same as the intrepid colonisers back in Palaeozoic times: arachnids, springtails, plus a range of insect groups. A mere two months after Krakatau exploded, and long before any plants became established, more than 40 species of spiders had arrived by ballooning – drifting through the air on gossamer threads. A similar phenomenon has been documented in Hawaii where, on new lava flows, crickets and wolf spiders are among the first colonisers, well before plants.

Springtails, present in some of the earliest terrestrial fossil deposits, like Rhynie, are very early, opportunistic colonisers of new islands. On Krakatau, and on the volcanic island of Surtsey in the North Atlantic, these detritivores had appeared almost before the lava had cooled. On these islands early arachnids and springtails rapidly established an aeolian ecosystem, existing on other wind-borne arthropods, either by scavenging or by direct predation. So, who knows, maybe the tenacity of arthropods in establishing ecosystems in places on Earth today where no plants exist points to the earliest terrestrial ecosystems having evolved in a very similar fashion.

6

Jaw Wars: the Empire Bites Back

The evolution of jawed fishes

Everyone has horrible memories of going to the dentist. As a four-year-old I remember vividly having a molar taken out without anaesthetic simply because the dentist offered me a gigantic needle and, not knowing any better, I declined. This put me off going to the dentist for some 14 years, and then I required lots of dental work, but miraculously I still have a reasonably good set of teeth (or what's left of them) nowadays. Teeth, though, have always been a pain in my life, and, like most people, except those who are masochistically inclined, I still squirm at the thought of having to visit my dentist, nice as he really is. So, where did this legacy of teeth begin? The answer lies back at least 400 million years ago. Following the rise of the great jawless wonders, the agnathans, the stage was set to roll for the greatest battle of lower vertebrate supremacy – jaws or no jaws, an eye for an eye and a tooth for a tooth! Surprisingly, there is a vital link between jaws, eyes and teeth, which will be revealed later in this chapter.

The first true fishes, known affectionately as 'agnathans' take their name from the Greek meaning 'without jaws' as they are a group of mostly extinct fishes that lacked true bony jaws and teeth. Today, the parasitic eel-like lampreys and scavenging hagfishes are our only representatives of this once-flourishing group which reached an acme of diversity in the Early Devonian, about 400 million years ago. Although modern agnathans may seem to be boring creatures of little interest to most humans, apart from being a very minor food source in some countries, the study of fossil agnathans is vital to our understanding of many important anatomical transformations that took place early in the history of vertebrates. The fossil agnathans are our only window into understanding the origins of jaws, the evolution of cellular bone, and the complete organisation of the standard vertebrate head pattern. Their significance to zoology therefore quickly becomes clear. This aside, it is the simple beauty and mystery of the numerous, bizarre-looking agnathan fossils that interests us, as much as their quintessential scientific value.

Lampreys, incidentally, once played a minuscule role in recent human history. King Henry I of England died in Lyons, France, on the 1 December 1135, after an excessive banquet in which he gorged himself full of lampreys. They are still considered a great delicacy in some parts of Europe. They are not something that you are likely to be able to pick up in your local supermarket along with a tin of baked beans, though. Some lampreys are parasitic, feeding on other live fishes by attaching themselves with an oral sucker disc. Other lampreys forgo this Dracula-like existence and lead a much more boring life buried up to their necks in the muddy bottom of rivers. The relatives of the lampreys, the myxines, or hagfishes, are far more primitive in many aspects of anatomy, and are deep-sea carrion feeders. Hagfishes are the largest living jawless fishes, growing up to 1.4 metres in length.

Lampreys and hagfishes have fossil records extending back to the Carboniferous Period. They have remained almost unchanged throughout the past 340 million years. The extinct fossil agnathans include six major types of armoured and some non-armoured forms, most of which had evolved by the start of the Silurian Period, about 430 million years ago[1]. However, the first jawless fishes include the primitive Ordovician forms from Australia (*Arandaspis* and relatives), South America (*Sacabambaspis*) and North America (*Astraspis* and *Eryptychius*). These all share the primitive feature of having numerous paired openings for the gills, a feature reduced in number in all subsequent jawless fishes except for some anaspids. The bone making up the shields of the North American forms such as *Astraspis* is composed of four layers of phosphatic minerals including fluorapatite and hydroxylapatite, with dentine-like tissues forming the hard outer layers, similar to our teeth. This suggests a close relationship to the heterostracans, a diverse group that had similar shields to these Ordovician forms but which possessed only one pair of openings over the gill chamber.

The earliest relatively complete fish fossils come from Alice Springs in central Australia, where they occur in fine-grained sandstones dated at about 470 million years old, as revealed in Chapter 4. When the fossils were first found in the rocks in the mid-1960s the strata were immediately thought to be much younger, namely Devonian, in age, because at that time Ordovician fish fossils were virtually unheard of. Further collecting at the sites by Alex Ritchie of the Australian Museum in the 1970s and 1980s has yielded a number of good specimens of these early primitive fishes[2]. The shields of *Arandaspis*, named after the Aranda Aboriginal people, are not preserved as bone but as impressions in the ancient sandstones. These tell us exactly what the shape of these armoured agnathans and their body scales were like. *Arandaspis* (see illustration on page 55) has a simple dorsal (top) and ventral (belly) shield with up to 14 or more paired branchial plates covering the gills. The eyes were tiny and situated right at the front of the head, like the headlights on a car, and there were two tiny pineal openings on the top of the dorsal shield. These probably functioned as light-sensory organs. The tail is largely unknown, except for the fact that it bore many rows of long trunk scales, each ornamented by many fine parallel ridges of bone, making them comb-shaped. *Arandaspis* occurs with a number of previously unknown forms of jawless fishes, most of which are still being studied by Ritchie.

The discovery of the world's first complete Ordovician fish fossils in central Bolivia by French-Canadian student Pierre-Yves Gagnier in the mid-1980s caused wide scientific interest when preliminary results were first revealed to the general world in *National Geographic Research* magazine in 1989[3]. These fish, called *Sacabambaspis* after the town of Sacabambilla in Bolivia, are slightly younger than the Australian fossils (around 450 million years old), but much better preserved. They show the entire articulated armour and body form of the fish (see illustration on page 55). Like *Arandaspis, Sacabambaspis* has a large dorsal and ventral shield with numerous rectangular branchial plates, small eyes at the front of the skull and paired pineal openings. It also has a rounded plate on each side near the front of the head. The body is covered with many fine elongated scales, and although it lacks paired or median fins, the tail was quite well developed.

The North American fishes *Astraspis* and *Eryptychius* were long known as the earliest fish fossils, their abundant remains coming from the 440 million-year-old Harding Sandstone of Colorado, preserved as isolated small fragments of bone. Only recently has an almost-complete fish been found. This shows that the overall body of *Eryptychius* looked like *Sacabambaspis*, although it has much coarser, rounded scales covering the tail, and its shield is made up of many polygonal units, called tesserae[4]. The significance of these fossils lies in their excellent bone preservation, and they have played an important role in the discovery of how primitive bone evolved. The bone of *Astraspis* has four layers: an outer thin layer of enameloid capping a second layer of dentine, which forms the ridges and tubercles of the plates, a third layer of cancellous or spongy bone, and a fourth basal layer of aspidin, a layered hard tissue that lacks bone cells.

Although these primitive Ordovician agnathans closely resemble the Heterostraci (one of the major Silurian and Devonian groups of agnathans), they lack one distinct feature of that group – a single external branchial opening for the gills. In many other respects, such as having a shield formed of numerous plates, and similar bone structure, they are very similar. The heterostracans most likely evolved from such ancestral stock. Yet despite the huge radiation of the many types of jawless fishes in the Silurian and Devonian, the next major step in vertebrate evolution was to be the appearance of jaws and teeth, suddenly putting the business end of the fish back into the limelight. As fish gathered momentum, and reached relatively advanced levels of locomotory skills, it was the feeding mechanisms that were set to explode, allowing a platform for widespread experimentation and niche occupations. The War of the Jaw was ready to rage.

The rise of the first jawed vertebrates (called 'gnathostomes') from jawless fishes is one of the great unsolved problems in vertebrate evolution. The classic interpretation of how jaws first originated in fishes is by modification of the front gill arch support bones. As these bones supported the mouth, and dermal scales (then having a tooth-like structure and shape) invaded the mouth, the first primitive set of jaws and teeth would have been formed. Supporting this view is the fact that primitive jawless fishes have many more pairs of gill arches than the jawed fishes, some having up to 20 or more paired gill pouches. Furthermore the dermal scales of most bony agnathans have dentinous tissues underneath the enameloid crown, with a bony base, some also having a pulp cavity (for example, the thelodonts). Thus the earliest scales of these fishes are practically teeth.

However, despite the simple scenario for the physical formation of jaws, the mechanism of how this may have come about has eluded scientists for some time. I have a theory about this which came to me in 1992 while I was listening to a lecture by Brian Hall, of Dalhousie University in Canada, one of the world's eminent developmental biologists, while he was visiting the Anatomy Department at the University of Western Australia. I was listening to Brian talking about chicken embryos, more specifically how some bones are formed by reactions with other tissue types setting them into developmental action. For example, the lower jaw (Meckel's cartilage) is kick-started by an interaction of the matrix around the scleral cartilage (the cartilage framework around the eye). It's called in technical jargon 'a matrix-mediated epithelial–mesenchymal interaction'! What this really means is that the formation of the lower jaw in the chick embryo is started by the cartilage buds of the eye making contact with the embryonic matrix.

Thus an eye for an eye, a tooth for a tooth! Suddenly a light went on inside my head as I searched the dusty corridors of my memory for evidence of similar processes in fossil fishes. Eureka! Looking at the fossil record of jawless fishes we find that only one group, the osteostracans, have ossified sclerotic bones around the eyeball. This suggests that only they, or their immediate ancestors, could have had the developmental potential to develop jaws from the same processes that initiate lower jaw cartilage formation (assuming the processes of jaw formation that we see in the chicken are universal for all jawed vertebrates).

This idea reinforces the recent work by Philippe Janvier of the Museum National d'Histoire Naturelle in Paris and Peter Forey of the Natural History Museum in London, which shows that the osteostracans share more advanced features in common with jawed fishes than with any other agnathan group, such as perichondral bone enveloping the braincase (a kind of thin laminar bone that enveloped cartilage)[5].

The inside surface of the head shields of osteostracans indicates that the first gill arches were situated well forward of the eyes and directly above the mouth. Osteostracans presumably had cartilaginous gill arch supports as shown by impressions of gill structures on the visceral surface of their shield. Therefore the evolution of the anterior gill arch elements to become primitive jaws did not necessarily involve large-scale structural reorganisation. The hole for the mandibular and maxillary nerves is visible on the first branchial ridge in the fossil head shield of *Scolenaspis* and in several other osteostracans. The developmental link between jaws and scleral bone formation, as discussed above, is further emphasised by the fact that the ophthalmic nerve is also a branch of the fifth cranial nerve (as are the mandibular and maxillary nerves), serving to innervate the eyeball with small ciliary nerves.

The reason why jaws evolved may not be simply related to the functional significance of the structures themselves (for example to support teeth and improve feeding ability) but may be primarily under developmental control influenced by the timing of neural crest cell migration. These are cells that arise from the developing spinal cord. While these cells play no role in forming skeletal tissues in the trunk, they do play a role in skeletal tissue formation in the head. Many early agnathans had well-developed oral plates lining the ventral border of the mouth giving them an effective feeding mechanism that operated in a manner similar to a lower jaw. Some osteostracans, such as *Tremataspis*, had an anterior median lamina covered by denticles situated inside the roof of the mouth, indicating that ectoderm tissue invaded the buccal cavity (perhaps derived from the neural crest). This implies that a structure may have acted against this surface for food reduction, possibly a denticle-covered tongue or rasping organ[6].

Thus jaws probably originated from modification of the front gill support bones, and teeth invaded the mouth cavity primarily as pointed scales. Some primitive jawless fishes like thelodonts have recently been shown to have had areas of bony scales inside the mouth to help reduce prey or direct the flow of water currents inside the mouth and gill chamber. The most primitive jawed fishes may well be the sharks whose earliest scales actually resemble those of the most primitive thelodonts. New discoveries of well-preserved complete thelodonts from Canada show preservation of the outline of their stomachs, a feature thought to be absent from agnathan fishes[7]. The presence of a well-developed digestive system implies that they were more advanced in their feeding habits and food reduction mechanisms than previously thought, and has given rise to suggestions that the thelodonts may even be closer to the first jawed fishes than the osteostracans.

The appearance of jaws in the fossil record would naturally herald the abundant appearance of teeth at the same time. Sharks, for example, have hundreds of teeth in their mouths, and because they grow and shed them continuously through life, the average shark may shed something like 20 000 teeth, which ultimately accumulate in the sediments. The skin of sharks contains many thousands of tiny placoid scales, which are also shed into the sediment after death. Yet there is a strange enigma about the fossil record of sharks which hints at an unusual evolutionary

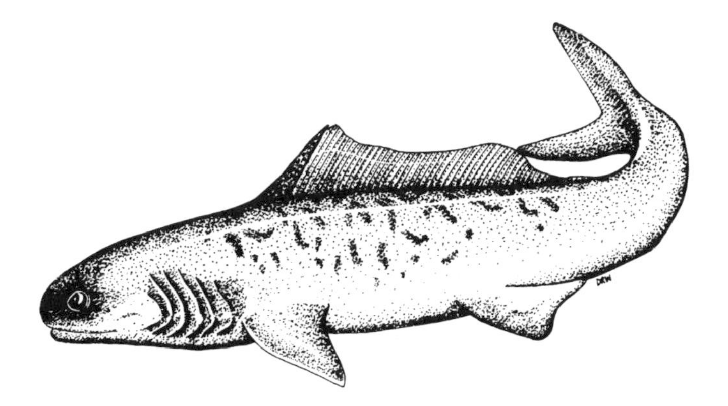

Jaws from the deepfreezer of time: an early fossil shark from Antarctica. This reconstruction represents the oldest fossil shark known from relatively complete remains, Antarctilamna, *dated at 380 million years old. Its teeth, scales and fin spines have now been found throughout Gondwana, in Australia, Antarctica, the Middle East, South Africa and South America. Sharks may well have originated in Gondwana.*

story. The fact is that the fossilised scales of sharks go a long way further back in time than their teeth. The oldest known shark scales are from the Late Ordovician–Early Silurian of Mongolia, yet the oldest known teeth, that are recognisable because they have a true root base and canals, are of Early Devonian age, some 30 million years later. As the numbers of teeth and scales are both high per individual shark, the absence of teeth in Silurian strata would seem to be a real phenomenon, as many thousands of samples processed around the world for conodonts and other microfossils would have surely turned them up by now. So, this indicates that the first sharks had scales of recognisable morphology akin to modern shark scales, yet they lacked teeth. Perhaps then, the appearance of the first teeth is correlated with the evolution of jaws in sharks?

Sharks first rose to the fore as a group back in the Middle Devonian in Gondwana, where their teeth appear commonly in some deposits, representing a diverse fauna of shark types. The story of the discovery of some of these early shark faunas takes me back to my days in the icy wastelands of Antarctica. In late 1991 and early 1992 I was in Antarctica as part of a 'deep field' expedition, sledging through the Cook Mountains and Skeleton Névé regions studying the Devonian sedimentary rocks there. My job was to collect fish fossils which were abundant in layers of the 380 million-year-old Aztec Siltstone, an accumulation of sandstones, shales and mudstones from ancient freshwater streams and lakes. Despite the hazards of fierce storms, numb fingers and toes, almost falling down crevasses and having an avalanche dumped on me, I was almost continually on a high of discovery. Nearly every site we visited was a new fossil locality.

To my surprise we found a large number of fossilised sharks' teeth in these deposits which were later described by Gavin Young and myself[8] as belonging to three new genera and species. However, the full total of the shark teeth species found in the Aztec Siltstone assemblage represents a large number for the Middle Devonian – at least six or seven species, and a dramatic increase in size, as older sharks' teeth from the Early Devonian are all quite small (2–4 mm). Some of the Aztec sharks had teeth 2 cm in height, indicating large predatory fishes maybe 3 metres long were cruising the ancient river ecosystems. These observations led me to suggest that Gondwana may well have the been the place where sharks not only first originated, but first radiated in terms of size and diversity before spreading out around the world.

By Early Carboniferous times some really large sharks were about, like *Edestus giganteus*, from North America, which had large serrated teeth up to 7–8 cm high set on a continuous whorl. Such megapredators would have reached 6 metres or more in length. The largest known fossil shark predator was *Carcharocles megalodon*, a distant ancestor of the mako shark lineage, whose huge teeth measure up to 18 cm. Conservative scientific estimates place the total length of this killing machine at 15 metres long. This beast lived in the seas of the Miocene and Pliocene Periods and was thought to have become extinct about 2 million years ago. I have suggested a theory in my book *The Rise of Fishes* that this shark may have died out when its prey, the great baleen whales, began regular migrations into the icy Antarctic seas, not only for feeding but also to escape the giant sharks which couldn't have tolerated such freezing waters.

So, although sharks are amongst the front runners for the first jawed fishes in terms of their primitive scales resembling teeth, the oldest fossil animals with proper jaws and teeth are actually the acanthodian fishes of Early Silurian age, often represented in the microscopic residues from dissolved limestone as isolated teeth, scales and fin spines. Acathodians have been likened to 'spiny sharks' in past literature although their affinities appear to lie closer to the higher jawed fishes, such as osteichthyans (true bony fishes) because of their scale structure and gross morphology of the braincase.

Still, the oldest acanthodians include forms with distinct gnathal bones that have strong teeth fused on to these bones. This group, called the ischnacanthids, consisted of moderately large predators in the Early Devonian seas, some reaching sizes of 2–3 metres, like *Xylacanthus grandis* from Spitzbergen. Unfortunately we have a poor overall knowledge of the group as no perfect, three-dimensionally preserved specimens have been found enabling us to study their internal ossifications. Only the last, and arguably most specialised, member of the acanthodians, *Acanthodes*, is known from relatively complete remains, including the braincase and gill arch skeleton. *Acanthodes* had lost its teeth, becoming secondarily adapted to filter feeding, so it sheds no useful information on the origins of jaw and teeth.

The most successful of all the early jawed fishes were undoubtedly the placoderms which appeared in the Early Silurian but really flourished during Devonian times, becoming extinct at the end of that period. They take their name from the Greek meaning 'plated skin', named because of their mosaic of bony armour plates that enveloped the heads and trunks of these fishes. They possessed shark-like tails and in their overall general anatomy were very shark-like. Some workers back in the 1980s, like Peter Forey and Brian Gardiner of the Natural History Museum, London, suggested that placoderms were closer to bony fishes than to sharks[9], but others have argued that they are more primitive and are more likely to be a split-off from the common ancestor of sharks and rays (chondrichthyans)[10].

I played a small role in this debate when in August 1986 I found a specimen of a placoderm at Gogo, in the Canning Basin in the north of Western Australia, that possessed annular cartilages preserved as bony ossifications. These structures form a ring around the nostrils and are only found in sharks today. For many years the Swedish scientist Erik Stensiö had argued from the shape of the nasal capsules in placoderms that they must have possessed annular cartilages, but he never saw one fossil showing such a feature. Thus to my surprise, the specimen from Gogo (subsequently named *Mcnamaraspis kaprios* after co-author Ken McNamara for his 40th birthday, and the Greek meaning 'boar-like', though not because of his ugly looks), actually showed the cartilages in position in front of the nasal capsules. In my paper describing *Mcnamaraspis*[11], I argue for a close link between placoderms and chondrichthyans, based on this and some other features newly revealed by the Gogo specimen. Yet the evidence is still quite equivocal as to where placoderm affinities really lie.

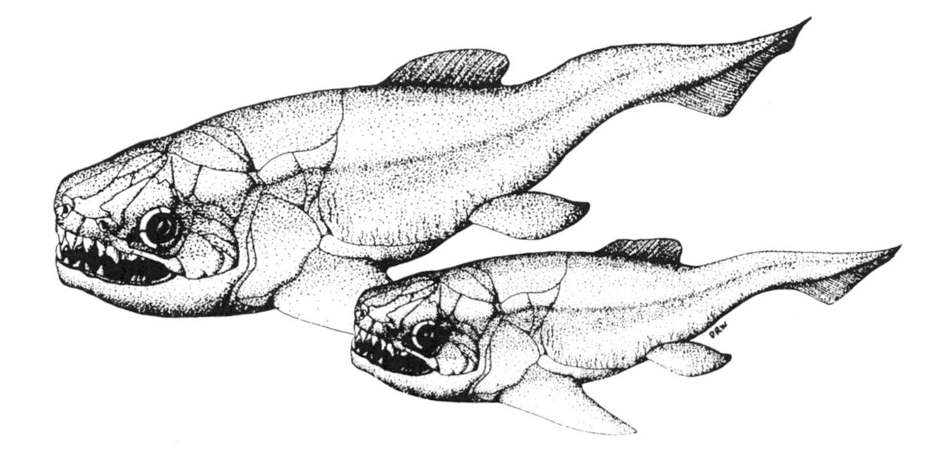

Anyone for a bite? This reconstruction shows two armour-plated placoderm fish, Mcnamaraspis, *which came from the fabulous Gogo fish site in the far north of Western Australia, dated at around 370 million years old. Placoderms were amongst the first fish to have teeth-like cusps on their jaws, but lacked true teeth that grew with roots. Some placoderms reached enormous sizes like* Gorgonichthys, *which may have sometimes been more than 6 metres long.* Mcnamaraspis *is not long. It fits snugly in its namesake's hand.*

Aside from this, what are the teeth of placoderms like and how do they relate to other jawed fishes? The teeth developed by placoderms are not real teeth, with roots set into a discrete jaw bone, like those of the osteichthyan fishes which eventually led to reptilian and mammalian teeth. Rather, they are well-developed pointed cusps that develop on the jaw bone itself. Thus we see placoderms having a multitude of dentitions, but these usually consist of rows of pointed cusps on gnathal bones, or clusters of small denticles or pointed cusps on sheets of bone, similar to the parasphenoid (the median denticulated palate bone underneath the braincase) of osteichthyan fishes. Incidentally, this is one strong feature that placoderms share with osteichthyans that is lacking in acanthodians.

The placoderms included the first of the really giant megapredators of the seas – huge carnivorous armour-plated fishes like *Gorgonichthys* and *Dunkleosteus*, whose lower jaw bones measure over a metre long and which were armed with sharp, trenchant cusps measuring some 20 cm in depth. These monsters would have easily reached sizes greater than today's great white sharks, and were clearly at the top of the food chain of their day. Yet, how could they survive covered in such heavy bony plates? And what did they eat? The open seas of the Late

Devonian also teemed with sharks, some up to 2 or 3 metres in length. Yet, from their lack of bone, and sleek, sporty body design, these fishes must have been fast-swimming creatures. It is hard to imagine the Godzilla-like *Dunkleosteus* actually catching one of the GTX racing sharks like *Cladoselache*, unless ambush tactics were used. It is far more likely that the megaplacoderms were hunting each other, as a high diversity of dinichthyid and related large placoderms coexisted in the same ecosystems and were fossilised in the same beds. For example, the Cleveland Shale in Ohio and New York State has yielded thousands of such specimens.

The placoderms included many unusual varieties, some of the most successful being the most bizarre that would probably have been voted by their high school graduation panel 'least likely to succeed'. Such an example is *Bothriolepis*, a pitiful-looking little thing with clumsy bony pectoral appendages and puny jaws (lacking teeth), small eyes and nostrils located centrally in the middle of its head shield. Its closest living counterpart, niche-wise, might well be some of the armoured freshwater catfishes of South America, like the *Plecostomus* group. They seem to have similar body shape with central eyes and nostrils, and well-developed pectoral fin spines similar to the spiny pectoral appendages of *Bothriolepis*. They use these arm-like structures to cling to rocks in fast-flowing streams while they graze off algae.Yet despite its apparently poor body design, somewhat like a box with a weak tail, *Bothriolepis* is perhaps the most common fossil fish found in the Devonian. It is known from every continent, from over 100 valid species (based on well-preserved, more or less complete fish), and from palaeolatitudes spanning from equatorial marine deposits, like Gogo in Western Australia, to near the Devonian South Pole in South Africa.

Back in the 1940s Robert Denison of the American Museum of Natural History serially sectioned a *Bothriolepis* and found a different type of sedimentary rock inside it. This had preserved the shape of the alimentary canal and Denison concluded, from examining its 'gut contents', that the fish was a mud grubber, receiving its nutrition from the ingested organic particles derived from the muds on the bottom of the ocean floor[12]. Denison also found something else inside the box-like shell of the *Bothriolepis* – a pair of sac-like organs, which he suggested were lungs. Also, it has been suggested that the peculiar mobile, bony pectoral appendages of *Bothriolepis* were used as props that enabled the fish to 'walk' itself between pools of water, or as aids in digging itself down deeper into the muds upon which it fed.

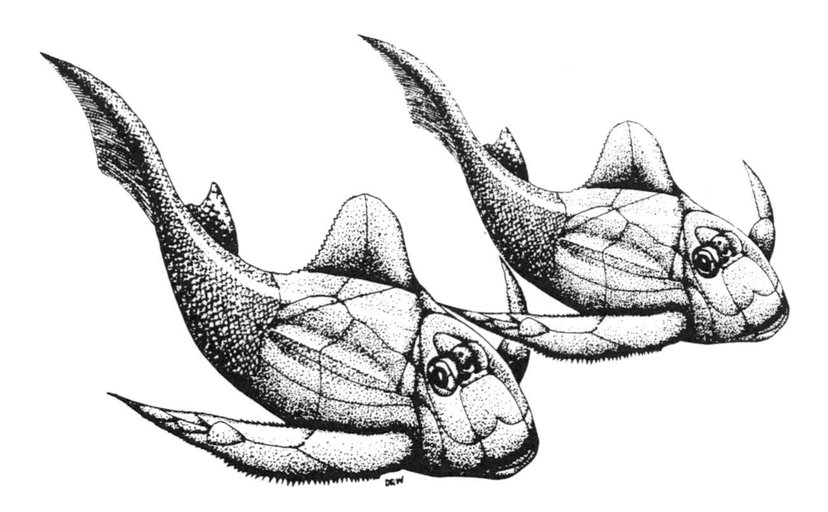

Maybe the most bizarre vertebrate to have ever evolved, the placoderm fish Bothriolepis *takes the cake for having its arms inside-out. With a body the size of a meat pie, a bony outer covering protected the muscles and tissues inside the pectoral appendage. Such 'arms' may have helped the fish drag themselves upon land for short adventures, or alternatively assisted in digging their little heads deeper into the organic-rich muds upon which they fed.*

Thus the picture emerges of *Bothriolepis* leaving one pool of water, breathing air for short periods, whilst dragging itself on its bony arms in search of new, uninhabited mud pools. This scenario would only work if, at the time, there were no large predators on land to chase and eat the ungainly fish as it hobbled from pool to pool; and that the organic muds were not rapidly ingested by the invertebrate faunas inhabiting those same environments – perhaps because the preponderance of jawed fishes were keeping them in line quite effectively. Even so there were some pretty large eurypterids (sea scorpions) about, marching around on the land at that time between pools of water, as shown by their trackways (see Chapter 5). It is quite likely that at least some of the *Bothriolepis* may have fallen prey to the voracious arthropods.

So, at this time, during the Late Devonian, major changes in the makeup of the fishes that swum in these ancient seas were about to occur. Placoderms were the most diverse and successful group of fishes, in all environments. They reached the largest sizes (up to 7 or 8 metres), had the greatest diversity of feeding mechanisms, and some, like *Bothriolepis* and its antiarch brethren, may even have conquered the land for short periods. So, why did their mighty empire suddenly come crashing down around them at the end of the Devonian? What happened to them that allowed the main jawed fishes of today to succeed them?

Firstly, we must look at which jawed fishes were around in the shadow of the placoderms – the sharks, the acanthodians and the bony fishes (osteichthyans). The last-mentioned consisted of three principal groups, the ray fins (actinopterygians) which today form more than 99% of all living fishes (for example, trout, salmon, goldfish, seahorse, eel), the lungfishes, with a few living examples in Gondwanan countries (Australia, South America and South Africa) and the crossopterygians, with one relict species, the coelacanth *Latimeria*. In the Devonian these ratios were turned around so that the greatest diversity of bony fishes were the lungfishes (about 45 Devonian genera) and crossopterygians (about 50 or more genera), with ray fins just emerging (under 10 genera). The ability to gulp air and thus transgress environmental boundaries only evolved within lungfish in the Middle Devonian, and became commonplace in the Late Devonian, thus perhaps giving them an edge on other groups[13]. Later forms had a further adaptation to survival: the ability to aestivate.

By Middle Devonian times some of the advanced crossopterygians appear to have developed the ability to breathe air. They possibly gave rise to the first amphibians by the early Late Devonian, if not sooner (see Chapter 8). Trackways found in eastern Australia point to the possibility of tetrapods walking around on land at the beginning of the Devonian Period[14]. Acanthodians were on the wane, with few forms surviving the end of the Devonian extinctions. As for sharks, they were steadily increasing in diversity throughout the Devonian Period (from less than 5 species in the Early Devonian to over 40 species by the Late Devonian).

Like the mysterious decline of the dinosaurs at the end of the Cretaceous Period, the end of the placoderm empire came about swiftly at the close of the Devonian Period. Many higher group extinctions occurred, so although placoderms were completely wiped out, so too were several higher taxonomic groups of other fishes (such as osteostracan and thelodont agnathans, onychodontiform and porolepiform crossopterygians, ischnacanthiform acanthodians and many families of lungfishes). Sharks appear to have sailed right through this time boundary unaffected, just as they did across the traumatic upheavals of the Cretaceous/Tertiary boundary[15].

The success of the sharks appears to lie in their teeth. These are the most changing and variable feature of the group, and, as fossils, their most diagnostic part. At the dawn of the Carboniferous Period a major radiation occurred within the chondricthyans, reflecting an ability to occupy rapidly all of the niches left over by the placoderms. There is no evidence of Late Devonian sharks already doing this alongside placoderms, just a sudden replacement of one group (placoderms) by another (the sharks and their kin). The first holocephalomorphs suddenly appeared (today represented by ratfishes and chimaerids), and many new groups of bizarre sharks emerged, some having pavement-like crushing dentitions. The teeth of sharks hold the key to their success – through increasingly complex histology and advanced tissue types that gave them great strength, as well as their ability to be replaced thoughout life. Thus an average predatory shark may have up to 600 or so teeth inside its mouth at any one time, but only the forward rows are effective in biting or seizing prey. However, during its lifetime it may shed over 20 000 teeth; thus it always has a sharp row of new teeth coming up to replace the worn or broken ones. No other group of fishes has this ability.

So, back to the mysterious decline of the placoderms. The answer to this mystery is far from clear. While there was an increase in fossil shark species during the Late Devonian, there is no dramatic decline in placoderms at this time, although some groups of placoderms seem to have dropped out by the Frasnian–Famennian extinction event in the midst of the Late Devonian. One can only postulate that placoderms settled into their niches and continued in their lifestyles, while sharks were exponentially increasing in species numbers (at least from the abundance of their teeth in the fossil record), and continually experimenting with new dentition designs. Perhaps they suddenly hit upon a formula allowing great evolutionary flexibility with their dentition, and this, coupled with their speedier and lighter body design, enabled them to manage suddenly a *coup d'état* and simply oust the placoderms in a rapidly short geological timespan.

Thus the evolution of teeth and jaws appears to have been one of the principal factors in shaping the future pattern of modern-day fish faunas. Since the beginning of the Carboniferous Period the balance of a high number of osteichthyan fishes (primarily actinopterygians), with sharks in high abundance, combined with a low ratio of crosspterygians and lungfishes, was also set in stone. Over the vast expanse of geological time between then and now the only changes have been in the fine tuning of their ratios and the numbers of species within each group.

All the while the Jaw Wars were raging in the seas, on land the first four-legged animals, the amphibians, slowly and almost imperceptibly emerged from the stock of advanced sarcopterygian fishes. But this, the greatest step in the evolutionary history of the backboned animals, is another story.

7

Tracking Down the Bugs

The evolution of insects

It was late in the afternoon. Another perfect winter's day in Western Australia: the sky such a deep blue you feel you could dive into it. The sun's rays were perfect now, raking low across the weathered sandstones that were outcropping intermittently high above the Murchison River gorge. Kris Brimmell of the Western Australian Museum was working her way along a goat track, many kilometres from the nearest vehicle track. Like all good geologists her head was down, scanning the ground for interesting-looking trace fossils. Although few and far between there is always the hope of finding just the perfect one.

It's hard to say what made her look at a small, scruffy piece of rock that would fit neatly into the palm of your hand. Whatever it was, her eye was caught by distinctive patterns on this particular rock. Had she been here an hour earlier, when the sun was higher in the sky, it is doubtful whether she would have noticed it.

'Good grief! It's a trilobite!' were her first thoughts. She bent down, picked it up and angled it to the Sun to pick up all the details. Here it was at last, after people had been looking for that enigmatic body fossil in the 420 million-year-old Tumblagooda Sandstone for decades – an actual real live fossil. Well , OK, not live, but a long dead one; you know what I mean. A fossil that you could believe, not just a few rows of holes in the rock. This one had a body, and segments and looked very much like a trilobite.

But a trilobite? Here, in rocks that were meant to have been laid down by a river? Trilobites are meant to have only lived in the sea. That night back at camp Kris showed it to Nigel Trewin, ace trace fossil hunter. He hadn't seen the likes of it before, despite years of traipsing over rocks like these not only in Australia, but far away in the north of Scotland. I was in Africa at the time, and unaware of the find until I got back a few weeks later.

'I've found you something special, Ken' were Kris's first words to me on my return. 'It's on your desk. It's a body fossil from the Tumblagooda!'.

My first thoughts were that here at last, surely, would be the eurypterid I had long been looking for. The animal that I was sure had made many of the trackways. And with a fossil like that we could do something about narrowing down the age of the rocks. Nigel and I felt the sands were deposited somewhere near the Silurian/Devonian boundary because of the close similarity of the trace fossils to ones from better dated sequences in Antarctica. However, the palaeomagicians – sorry – palaeomagneticians, who look at remnant palaeomagnetism in order to work out where the continents were in the past, and their age, think it is older.

So, when I first looked at the specimen it was with a mixture of feelings – one, obviously, of great excitement, because it was such a well-preserved body fossil found in such a coarse-grained rock, but this was tempered by the feeling of being a little disappointed. Not only was it not a eurypterid, but it was the likes of something I had never seen in my life before. What on Earth was it? What could it tell us about the age of the rocks and of the animals that were roaming these sand flats so long ago?

For the next few weeks I immersed myself in the scientific literature dealing with all the arthropods that had lived in Late Silurian to Early Devonian times. Lots of strange animals lived in those times, as attested to by their fossilised remains that had been collected and described over the last couple of centuries. But could I find anything remotely like Kris's find? Nothing at all. The next step was to forget about its age. Maybe we were wrong; perhaps the rocks were older. They couldn't have been much younger because of the age of overlying rocks. Now was the time to forget about the age aspect completely and to dig deep into the rich literature of fossil arthropods throughout their 540 million year history.

Once I did that, it didn't take too long to come up with a close match. There appeared little doubt that our mysterious animal was a euthycarcinoid. A euthycarciwhat? I hear you ask. That was my reaction. I have to admit that I had never heard of these animals. And that is despite the fact that they had been described as a completely separate superclass of arthropods, on par with the insects and myriapods (centipedes and the like). I suspect that few of my palaeontological colleagues had heard of them either.

Just what was it that made me so sure that euthycarcinoids were wandering along sand flats in Australia 420 million years ago, especially since the oldest known euthycarcinoid was meant to be in some rocks that were 120 million years younger? Well, they are pretty distinctive-looking beasts. Just imagine a friendly cockroach. Merely add a few more leg-bearing segments (another eight to be precise) and you would have an animal that looks not too dissimilar to the euthycarcinoid.

Resting in the palm of my hand, it is hard to imagine that when this animal came to its final resting place, its remains would in some way be preserved for much later comers in the land colonisation stakes to look at. Its head is quite small, and not easy to make out. But it has a long multisegmented thorax, which on the upper surface consists of five plates known as tergites. These represent basic limb-bearing segments that have fused together. These tergites just peek through from the impression of the 11 pairs of limbs that this beast had, and which gave away what it was. To imagine how the euthycarcinoid was preserved, just imagine taking a cockroach and pushing its underside down into something like wet sand, or maybe plasticine. What you would get is the impression of the underside of the animal, its legs and any part of the upper body that poked through. So it is with this specimen. Behind the 11 pairs of legs are a further five, much narrower segments of the abdomen, none of which have legs. This combination of 11 pairs of legs and a leg-free abdomen are characteristic of the euthycarcinoids. Because this specimen was so different from all others, with my colleague Nigel Trewin, I named it *Kalbarria brimmellae*, the species name honouring its finder, Kris Brimmell[1].

Euthycarcinoids are certainly one of the more obscure groups of arthropods and, but for the fact that, as I will discuss at some length, it has been suggested that they may be the ancestors of the insects, they would be consigned to the back of the palaeontological drawer. Exactly what they were has been a topic of some debate for years. This group of rare arthropods had previously been known only from Late Carboniferous (300 million years old) and Middle Triassic (230 million years old) strata in eastern France, the United States and eastern

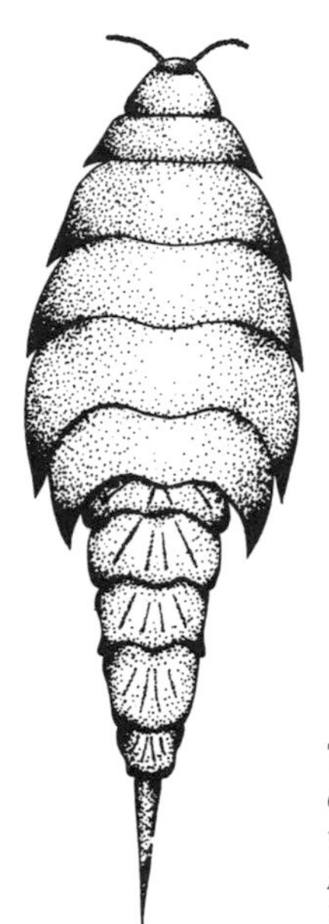

The mother of all cockroaches? This creature is possibly an early ancestor of the first insects, a euthycarcinoid, whose incredibly rare body impression was found in 420-million-year-old sandstones, near Kalbarri in Western Australia. It was about the length of a small hand, and is called Kalbarria brimmellae.

Australia. Early workers in the first part of this century thought they were crustaceans. In fact the name 'euthycarcinoid' literally means 'straight crab'. This view prevailed until the 1970s, when Fred Schram, now at the University of Amsterdam, suggested that they were merostomes[2]. The absence of uropods, characteristic features of crustaceans, plus their uniramian limb, led Jan Bergström of the Natural History Museum in Stockholm to conclude that they 'seem to show important similarities only with the uniramian groups and probably represent a distinct uniramian group comparable in rank with the Myriapoda and Hexapoda', in other words with centipedes, millipedes and insects[3]. Fred Schram and Ian Rolfe of the Museum of Scotland formalised this idea in their review of euthycarcinoids published in 1982[4]. They proposed the Euthycarcinoidea as a subphylum, or comparable rank to the Myriapoda, Hexapoda and Onychophora (velvet worms).

Prior to the Tumblagooda discovery, euthycarcinoids were thought to be exclusively freshwater. However, further discoveries in the Tumblagooda Sandstone suggest that, like a number of other arthropods living at this time (see Chapter 6), they were quite capable of walking out of water. Having found one body fossil I decided we had to go back to the site of the find and see if we could find the other half of the specimen, or maybe even some others. On reflection I should have realised that we wouldn't find the other half, because of the nature of the preservation. Like the trackways in the area this animal was preserved because it was caught out walking on a wet sand surface just as a huge dust storm came and overwhelmed it. While this would have preserved the impression of the animal's underside, there is no reason for the upper surface to have been preserved.

Even though we scoured the area for hours that day we found neither the counterpart, nor any more. However, an unexpected outcome was the discovery of tracks probably made by these very same animals. Because the site where Kris found the *Kalbarria* was rather remote, it meant we were in for a long walk. We set off at dawn and headed down to a narrow neck of the ridge high above the gorge, close to where the river meanders so tightly it almost strangles itself. Keeping to the edge of the cliff high above the gorge, we made our way up ragged steps of sandstones just as the first rays of sunlight hit the rocks. Again, fate was on our side. Had we come a little later, the higher angled sun would have revealed nothing. But its slanting, low angle early this morning raked a big slab perched at the top of the cliff above the gorge, illuminating a set of wonderful small tracks about three fingers' width wide. This multi-legged animal had obviously had problems walking along that distant day. Maybe it was a windy day, so it was 'crabbing' along at an angle. This meant that the repeated pattern that its many legs left made a clear impression showing exactly how many pairs of legs it had. You can probably imagine our delight when we counted them and came up with 11! The same number as the euthycarcinoid. The width of the tracks was much the same as the body fossil.

This meant that we not only had the remains of the euthycarcinoid body, but its trackway too. Like other trackways it must have been made on wet sand out of water – more evidence for this group of arthropods, even at a very early stage in their evolution, having the ability to be at least amphibious. But the story of our discovery of the life and times of the oldest known euthycarcinoids didn't stop there. Independently, Nigel Trewin had found, just a few hundred metres away, a bedding surface covered with some very curious impressions that looked as though somebody had got a small horseshoe and pushed it into the sand at an angle, curved end first. These impressions are perfectly euthycarcinoid shaped and may well have been its feeding traces. The bed of sandstone in which they are preserved also contains lots of tiny U-shaped burrows, the home, no doubt of some small worm or arthropod, upon which the euthycarcinoid might have been feeding.

So within the space of year not only had we discovered the existence of euthycarcinoids in these very old rocks for the first time, but we had also found their likely trackway, plus evidence of how they fed and probably what they were feeding upon!

Not only has the discovery of *Kalbarria* told us much more about how these rare, extinct arthropods lived, but it has helped substantiate a suggestion made early in the 1980s that euthycarcinoids may be a missing link in insect evolution. The higher level classification and interrelationships of these groups with each other, and with crustaceans, have been the subject of hot debate in the last few years. There are about as many ideas as there are people who proffer them. The concept of a group of 'uniramian' arthropods has come in for a lot of criticism in particular. However, some recent reviews[5] still consider that the hexapods and myriapods form a close grouping. With most of these reviews of arthropod higher level relationships, euthycarcinoids rarely get a mention, apart from works by Fred Schram. Yet I have picked up on a suggestion made by Schram and Rolfe to argue strongly that if we are to look for an ancestor for the insects we need look no further than the obscure euthycarcinoids.

To understand how an entire group of organisms can evolve a quite novel body plan it is necessary to look at details of the developmental history of the group, or of closely related groups. It is no good just looking at the adult of a species in one hand, and comparing it with the adult of another in the other hand, to unravel the evolutionary relationships between organisms. Evolution doesn't work like that. Evolution occurs by the entire developmental history (or ontogeny) of an organism, from its moment of conception, until it stops growing, changing from that of its ancestor. This can occur by different parts growing more, or growing less. They can grow more in a descendant than its ancestor by growing faster, or growing for longer. This can occur if growth starts relatively earlier than in the ancestor, or goes on for longer. By contrast a species, or parts of it, can grow less than its ancestor by growing slower, or growing for less time, either by starting later or finishing earlier. Don't be fooled into equating evolution with more complexity of form. Species can evolve more complex morphologies and behaviours, or they can equally well evolve less complex morphologies.

The role of changes to the degree of development (known as heterochrony) has been a much neglected aspect of evolutionary theory, but one which has come into its own again in recent years[6]. But for the purposes of our investigation into how insects evolved we need to look at one of these processes of heterochrony: one that involves a reduction in the amount of growth in the descendant, compared with its ancestor.

As long ago as the 1820s the great early nineteenth century embryologist Ernst von Baer remarked on the similarity between the early embryonic stages of some millipedes and centipedes (the myriapods) with adult insects. If this represents a true relationship, and is no mere will o' the wisp coincidental phenomenon, then we can conjure up two alternative models for the evolution of these groups. In terms of complexity of segmentation, myriapods are more complex than insects because they have more body segments and more legs. As such, it could be argued that these multisegmented organisms evolved from the simpler hexapods. However, the millipedes and centipedes have a much longer evolutionary history than insects, as the fossil record indicates. Dick Robison described the earliest myriapods from marine sediments that are about 500 million years old[7]. The earliest insect remains found in the fossil record are less than 400 million years old. This is the early Devonian collembolan *Rhyniella* from the Rhynie Chert in Scotland[8]. For the older myriapods to have given rise to the geologically younger insects, it is necessary to argue that the evolution of insects has occurred by a loss of body segments, combined with a loss of appendages, presumably from some ancestral centipede-like ancestor. However, rather than talking about any sort of loss, it is really better, and more appropriate I think, to regard the ancestral centipede or millipede as having given rise to a descendant by producing fewer body segments and appendages. Nothing was lost. It was never there in the first place! Insects, then, would be centipedes that failed to grow up.

How this could happen can be seen from looking at how living millipedes develop, from the egg to the full-grown adult. Upon hatching millipedes are not miniature adults, carrying beneath their bodies, like rows of umbrellas slung from beneath a roof, an overwhelmingly large number of legs. No, these first hatchlings enter the world equipped with just three pairs of legs. This is really the minimum a land-dwelling arthropod can cope with. The nature of the way insects, for instance, walk is that they have what is known as a 'tripod' gait. This means that at any one time there must be two legs on one side of the body and one on the other side in contact with the ground surface. This makes for a very stable gait, making it unlikely that the animal will fall over in the gentlest puff of wind. Any fewer than three pair of legs makes locomotion with a tripod gait impossible. The tiny millipede, while having a thorax bearing three pairs of appendages, was followed by an abdomen of 11 segments, to which were attached very reduced legs. As the millipede grows and passes through a number of moults, it increases its numbers of segments and limbs. The obvious conclusion to draw from this is that the evolution of insects from such a millipede or centipede ancestor would have occurred by some distant myriapod failing to grow up, becoming sexually mature, and stopping growing when in a larval state, with just three pairs of walking appendages. This retention of ancestral juvenile features in adult descendants is known as 'paedomorphosis' (literally 'child-shape').

This has been relatively widely accepted as a mechanism for insect evolution since the last century. But in their 1982 paper reviewing euthycarcinoids, Schram and Rolfe made a simple statement in a section that they entitled 'Phylogenetic speculation', that threw open another, quite different, interpretation. They wrote: 'if the euthycarcinoids could be shown to be paedomorphic derivatives of the sottyxerxids, then the dramatic reduction of limbs noted between these two forms during ontogeny, or phylogeny, might be regarded as an interim stage en route to hexapody'. In other words, insects might have evolved from euthycarcinoids. If so, how could this have happened? The euthycarcinoids and sottyxerxids are the two families that comprise the Euthycarcinoidea. Sottyxerxids were much more centipede-like, having over 30 pairs of legs, and 14 to 15 tergites (dorsal segments), but still having an abdomen behind it with no limbs. Euthycarcinoids, like *Kalbarria*, had just 11 pairs of appendages, 5 or 6 tergites, plus the abdomen. What euthycarcinoids and insects have that centipedes don't is an abdomen. Simply in terms of the numbers of segments, number of limbs and presence of an abdomen, we can construct a nice evolutionary trend from a centipede-like ancestor, bristling with dozens of segments and dozens of pairs of limbs, but no abdomen. This would produce a sottyxerxid-like descendant, by the posterior-most segments failing to sprout limbs during development. The same number of segments, but fewer legs. Later would evolve a euthycarcinoid, which produced fewer segments, even fewer pairs of legs (11); then finally a tiny descendant would evolve in which this trend is continued – even fewer segments, even fewer pairs of limbs (just three) and a limb-free abdomen. This earliest insect would have lacked wings – they came later, as I discuss below.

In such a scenario the euthycarcinoids become the logical missing link between myriapods and hexapods. But why has this model not been thought of before, and why wasn't it taken up by other workers after

1982? The main reason is a pretty obvious one in some ways. The geological record, as I have said, shows centipedes evolving long before insects. Until *Kalbarria* was picked up by Kris Brimmell, the earliest insect occurred in the fossil record a long time before the earliest euthycarcinoid, some 120 million years to be exact. With that one find, which showed that euthycarcinoids were scuttling around more than 30 million years before the first insect, euthycarcinoids can take centre stage in the debate over the evolution of insects.

While today our concept of what constitutes an insect is very much tempered by our relationship with the flies that annoy us at barbecues, or the mosquitoes that drive us mad at night as they whine around our heads like flying fretful children, it actually took some 50 million years for the first insects to take to the air. The fossil record indicates that the first insect wing evolved during the Late Carboniferous Period, about 330 million years ago. Surprisingly, we are again faced with the first in a lineage being the biggest. Some of the earliest winged insects were the largest of their kind ever. Forests of giant horsetails and ferns no doubt buzzed with the loud drone of giant dragonflies having wingspans nearly 70 cm across cruising the skies. Unfortunately the fossil record provides no hint as to what the first flying insects would have looked like. Presumably their wings were small and had originally been adapted for some other purpose. The question is, how and why did wings evolve in insects? If they evolved from some other structure, what was the pressure that induced a transformation in function? Recently, attempts have been made to address this problem from two directions. The first is just what environmental factors might have selected for a structure to become a wing. The second centres on the burgeoning research in genetics that is unravelling the developmental mechanisms that produced everything from the butterfly's flying work of art, to the mosquito's whining madness generator.

Two basic ideas about where insect wings came from have been doing the rounds for some years. One involves the idea that they originally functioned as gills in an aquatic ancestor, having been a modification of limb branches. Others have argued that wings arose not from a pre-existing structure, but were novel structures. Recently Michalis Averof and Stephen Cohen from the European Molecular Biology Laboratory in Heidelberg, Germany, have shown how the presence of two genes that have a wing-specific function in insects (genes called *pdm* (*nubbin*) and *apterous*) also occur in crustaceans[9]. This indicates that the embryonic origin of both insect wings and legs is comparable to the ventral and dorsal branches of crustacean limbs. Averof and Cohen therefore support the idea that insect wings evolved from a gill-like appendage, and suggest that the wings of insects are therefore homologous to the epipodites of crustacean limbs. Epipodites are outgrowths from the main limb in crustaceans which have respiratory and osmoregulatory functions.

Yet ideas have been speculative on how such structures, wherever they came from, could have been transformed into active, beating wings, that could support the weight of the insects. Some light has recently been shed on this question by James Marden and Melissa Kramer at Pennsylvania State University, on the basis of their studies of stoneflies from the Adirondacks[10]. Stoneflies' wings are too weak to allow them to fly, but they use them like sails on a windsurfer. Being such an ancient group, with fossil remains having been found in Carboniferous rocks, they are useful for formulating models of how insect flight evolved. Marden and Kramer found that by experimentally manipulating the size of the stoneflies' wings, the larger the wings, the faster they could skim across the surface of the water. Perhaps this is how wings may have first been used for locomotion, as intermediate structures derived from structures that were originally used for aquatic respiration. Their first use as tools of locomotion could possibly have been as oars to row across the surface of the water. As they got bigger, they could have been used more effectively as sails. If the rowing was muscle-assisted, then the bigger wing would have necessitated a larger muscle, big enough, eventually, for them to become wings capable of enabling the insect to escape the water entirely and take to the skies.

This idea is not based solely on stoneflies. Their relatives, mayflies, offer some support for this model. The particular stonefly that Marden and Kramer studied was called *Taeniopteryx burksi*. These insects have minute hairs on the legs and wings of both adults and juveniles. These hairs are useful when the insect is floating on the water, as they help stop the insect sinking. However, they do not help much in flying as they cause too much drag. Juvenile mayflies are also aquatic. They likewise possess hairy legs and wings and skim across the water. Adult mayflies, on the other hand have smooth, hairless legs and wings and, unlike, the stoneflies, are capable of flight.

In recent years a lot of experimental work has been carried out with the aim of trying to unravel the fundamental mechanisms that lie behind the development of insect wings. This research has centred on trying to elucidate the mechanisms of cell growth, particularly in the fruit fly *Drosophila*, and how the cells are arranged in the wing as it develops. This research on insects, not only on wings, but on the body as a whole, is showing that during development there are innumerable steps in the formation of structures like wings. Moreover, each step is critical to the next step. Any slight perturbation, either a delay or a premature initiation, can cause a profound disruption to the orderly sequence of events. Despite most 'programming errors' being fatal, occasionally the mistakes if they occur in the right place at the right time can prove advantageous and be preferentially selected.

Such may have been the case with insect wings. The first winged insects bore wings on every thoracic and abdominal segment, not only on one or two thoracic segments as modern insects do. Sean Carroll and his colleagues from the University of Wisconsin at Madison have argued that the timing of expression of homeotic genes (these are genes that play a critical role in triggering the genetic pathway that establishes the identity of the body segments in the embryo) has played a central role in determining wing number[11]. Yet, instead of contributing to wing formation, homeotic genes inhibit it. Carroll and his colleagues consider that during their evolution the role of homeotic genes increased. This is reflected in the reduction by repression of the growth of wings on a number of segments. When these genes are expressed can determine whether or not wings are produced. For instance, there is one homeotic gene that, if it is expressed in the embryo, inhibits wing development, but if it is delayed until later in ontogeny it has no effect at all.

While much of our understanding of the origins of insects in years to come will be derived from further studies of the developmental genetics of these and other arthropods, rare fossil finds, like *Kalbarria*, still have the potential to play an important role in helping us understand the ultimate origins of this epitome of biodiversity.

8

First Foot Forward

The vertebrates' escape from the sea

It is late in the day on an August afternoon in 1986. I am standing in Bugle Gap, about 100 km east of Fitzroy Crossing in the northern part of Western Australia. Around me rises the magnificent remains of an ancient fossil reef, that 370 million years ago teemed with a myriad of primaeval life forms. Today, after millions of years of constant erosion has removed the surrounding rocks, the core of the reef and all its associated sediments are exposed once again. It is just as if the sea has drained away, and what was once a sea of water is now a sea of spinifex grass. The land between the jagged limestone reefs represents the quiet back reefs and inter-reef bays of shallow seaways, some about 100 metres deep, where a coruscation of primitive fishes once swam. And while these tropical reefs of northern Australia flourished, far away on the other side of the world in part of what is now Scotland, the first land animals, little more than fishes with 'limbs', were clumsily plodding out of the rivers for the very first time and invading the land.

The ground around me is littered with hundreds of rounded limestone concretions which I smash with my geological hammer in the hope of finding fish fossils. The specimens from these concretions, which form part of a rock unit known as the Gogo Formation, are world famous for their exquisitely preserved, three-dimensional fish fossils. Suddenly, the hammer strikes and reveals a dazzling spectacle – a gleaming row of sparkling teeth jutting from a long jaw bone of glistening shiny white enameloid tissue. This tissue, known as 'cosmine', means that it could only belong to either a lungfish or a crossopterygian, both ancient lobe-finned fishes[1]. However, the pointed teeth along the jawbone indicates that it has to be a crossopterygian, but something different from the commonly found Gogo *Onychodus*. 'Crossopterygian', I should add, is an old-fashioned, nearly defunct term for a hotchpotch of various predatory lobe-finned fishes including forms like the living coelacanth *Latimeria*, as well as many extinct groups without vernacular names, such as osteolepiforms, porolepiforms and onychodontiforms.

Anyhow, the discovery of this jaw signalled that I had just found the first ever complete skull of an osteolepiform fish from Gogo, and just the thought of all that incredible detail on every bone, and the braincase preserved in perfect three-dimensional form inside the rock, made me go weak at the knees. What made this an especially memorable day was that literally only minutes before I had found another well-preserved fish at the same site, that at the time seemed to be just another run-of-the-mill armoured placoderm. When this specimen was prepared back in the lab, it too turned out to be something completely new to science. It was later destined to play a small role in Australian history when, on 5 December 1995, that fossil, which I had named *Mcnamaraspis kaprios*, was proclaimed by the Governor of Western Australia as the State Fossil Emblem of Western Australia. Australia's first fossil emblem.

The real find of that fateful day, though, was the crossopterygian skull. It turned out to be a specimen of *Gogonasus*, a genus which I had named in 1985 after the study of a single snout belonging to an osteolepiform fish[2]. The name actually means 'nose from Gogo' because that was all we had of it at the time. Not long after finding the new specimen I prepared it by dissolving the limestone in dilute acetic acid. This revealed an almost perfect skull, complete with braincase and gill arch bones, and confirmed its identification as *Gogonasus andrewsae*.

At that time, in the mid-1980s, a raging controversy was going on, between advocates of the lungfish theory (that lungfishes were more closely allied to tetrapods, or land animals) and the traditionalists who supported the origin of tetrapods from crossopterygian fishes. In a paper published in 1981, the late Donn Rosen, of the American Museum of Natural History, along with Colin Patterson, Brian Gardiner and Peter Forey, colleagues at the Natural History Museum, London, argued that the excellent preservation of the Gogo lungfishes revealed that one long-snouted form, *Griphognathus*, had an opening in the palate similar to the internal nostril of tetrapods (called the 'choana')[3]. Furthermore, they found evidence of a close relationship between lungfishes and living amphibians in their soft-tissue anatomy, evidence that was not present in the coelacanth and that could not be corroborated with extinct crossopterygian fishes. The main thrust of their argument was that the internal palatal nostril in osteolepiform fishes was not a proper nostril at all, but merely an opening in the roof of the mouth to accommodate the large fangs of the coronoid series of the lower jaw. Without it, they argued, the poor fish would stab itself in the mouth every time it ate.

Because at the time there were no three-dimensionally preserved osteolepiform skulls against which to test the theory, the debate raged on, based on different interpretations of crushed fossils from elsewhere, and other data not testable on the limited fossil evidence at hand. Erik Jarvik of the Natural History Museum in Stockholm has spent a lifetime describing the anatomy of the osteolepiform fish *Eusthenopteron* in great detail[4], and had shown clearly from his accurate reconstructions, based on a serially sectioned skull, that the choanae in the palate were nowhere near the lower jaw fangs. Yet it seemed as if the description of the beautiful Gogo lungfish skulls by Roger Miles in 1977 had flooded the market with such an overwhelming wealth of new, detailed anatomy, that the traditional views espoused by Jarvik had become irrelevant to understanding fish–tetrapod relationships. So late in 1986, after I had prepared the *Gogonasus* skull out of the rock, my first aim was to try to fit the lower jaws on to the palate to see if the lower jaw fangs did indeed fit into the supposed internal palatal nostril or choana. By matching the irregular line of the gape, a precise fit of the lower jaws to upper jaws and palate was possible, and by observing the closed mouth from the inside, it was clear beyond any doubts that the fangs of the lower jaw closed well away from the choana, which was open to grooves running along the length of the lower jaw. The choana was a nostril after all.

The result was so exciting that I published preliminary results in the *Australian Journal of Science* in 1987 followed by a full-colour report and picture of the palate and lower jaw fit in the *National Geographic Research* magazine in late 1988[5]. The significance of the find was reported further by Per Ahlberg of the Natural History Museum, London, in a brief report in *Nature* in which he concluded 'The Gogo fossil fishes provide such information [in reference to how new data to solve problems in evolution can only come from superb new specimens]; their scientific value is incalculable'.

Following the 1986 discovery of the first complete *Gogonasus* skull, a second skull was found on a joint expedition of the Australian National University and the Western Australian Museum in July 1990, by Dick Barwick of the Australian National University. It too is a complete, perfect specimen, and our joint description, with Ken Campbell of the Australian National University, of these three known specimens was published in 1997[6]. However, in the last 10 years since the first *Gogonasus* skull was found, a most remarkable series of new finds around the world has greatly clarified this grand step in evolution, when backboned animals left the waters to establish a stronghold on Terra Firma and confirmed crossopterygians as the precursors of tetrapods.

Prior to 1975 only two fossil amphibians were known from the Devonian Period: *Ichthyostega* and *Acanthostega*, both from East Greenland, and used extensively by Erik Jarvik to demonstrate their link with osteolepiform fishes. Fossil trackways found in Victoria, Australia, in the 1970s, also provided some evidence that the group was at least widespread by the Late Devonian. In 1984, a fossilised amphibian trunk with girdles and forelimb was found at the Andreyevka site in Russia, and named *Tulerpeton*. Surprisingly it had six digits on the hand, rather than the conventional pattern of five, as seen on most tetrapod hands and feet[7].

Then in 1977 a lower jaw bone was described controversially as belonging to an early amphibian. The fossil had been found in central New South Wales, Australia, and its identification as an amphibian[8] by Ken Campbell and Maurice Bell, also of the Australian National University, was immediately doubted by some overseas workers[9]. Named *Metaxygnathus*, it was thought to be slightly older than the East Greenland amphibians, being Late Devonian (about 370 million years) in age. If the interpretation was correct, not only would it have represented the world's oldest tetrapod body fossil, but it would furthermore have implied a possible origin for tetrapods at the other side of the world in Gondwana.

Metaxygnathus was quickly debunked as an amphibian jaw by Hans-Peter Schultze of the University of Kansas[9]. He argued that the lower jaws of the advanced lobe-finned panderichthyid fishes were so much like those of primitive amphibians that it was difficult to determine the fossil's affinities from lower jaws alone. *Metaxygnathus*, he argued, was probably not a tetrapod, but one of these fishes. However, when British fossil amphibian researcher Jenny Clack, from the University of

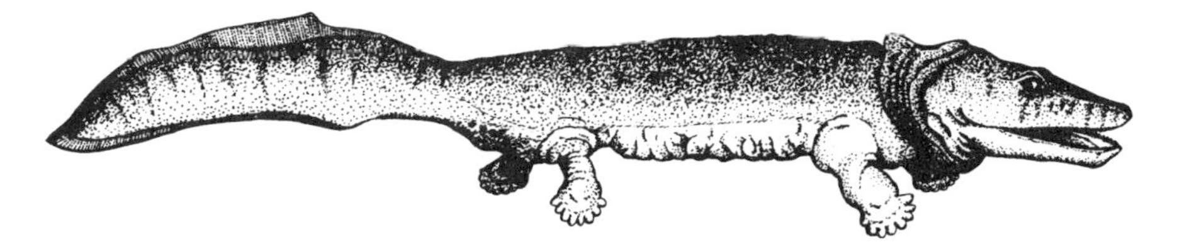

This somewhat Elizabethan-looking beast is actually a reconstruction of one of the world's earliest fossil tetrapods, the fox-sized Acanthostega, *from the Late Devonian of East Greenland, about 360 million years old. Note the many fingers and toes, the obvious gill slits behind the head and the fish-like tail (after the work of Mike Coates and Jenny Clack).*

Cambridge, looked at the specimen in Australia, she promptly reinstated *Metaxygnathus* as a good fossil amphibian jaw[10]. Further detailed study by Per Ahlberg of the Natural History Museum, London, and Oleg Lebedev of the Palaeontological Institute in Moscow, during a visit to Australia in 1995, left no doubts as to its tetrapod affinity. Yet as this single lower jaw was all that was known of the beast, little could be said about its evolutionary position amongst early amphibians at this stage. More on this one later as the story unfolds.

In 1987 a joint expedition to East Greenland, by members of the University of Cambridge and the Greenland Geological Survey based at Copenhagen, uncovered spectacular new finds of the first articulated skeletons of Devonian amphibians. Prior to this discovery, specimens collected in the 1930s by Danish and Swedish expeditions had been the basis of the first descriptions of Devonian tetrapods by Gunnar Save-Soderbergh, and later by Erik Jarvik, although only brief descriptions were published. The new finds, made by Jenny Clack, Per Ahlberg and colleagues, revealed such remarkable details about these first amphibians that a great deal has been learnt about their relationships and the evolutionary mechanisms involved in their evolution from fishes. So far only two fossil amphibians are known to exist in this fauna, which was dated as very latest Devonian (Late Famennian) in age (about 360 million years old). These are *Ichthyostega* and *Acanthostega*. Initially distinguished by differences such as skull roof patterns and shapes of cheek bones, the new material has revealed quite substantial differences in anatomy, and therefore probably lifestyles, between these two genera. The new material took years of painstakingly delicate preparation using fine dental drills and hand-held chisels. But it was to prove to be a gold mine of scientific discoveries. A series of papers by Jenny Clack and co-worker Mike Coates soon appeared in the scientific journal *Nature*.

The first such paper was published in 1989 by Jenny Clack[11], revealing that in one specimen of *Acanthostega* the stapes, or inner ear bone, was preserved intact *in situ*. The presence of a stapes was often seen by other workers as a prerequisite for the adoption of a terrestrial lifestyle. In advanced tetrapods it is associated with hearing, as the stapes supports a tympanum or eardrum. In *Acanthostega*, the stapes was still very much like that of the hyomandibular bone in the advanced lobe-finned osteolepiform fishes. Fishes, of course, do not hear in the same way as land animals like you or I, but use their hyomandibular bone (which was later to become the stapes in tetrapods) to brace the lower jaw joint and back end of the palate and to direct the hyomandibular nerve down to the palate. Clack found that the notch at the back of the skull in *Acanthostega*, which was thought to be for the tympanum (and connected by the stapes), was probably a primitive fish-like opening called a spiracular slit, and that the stapes in *Acanthostega* controlled palatal and spiracular movements. These movements were more vital to primeval air breathing than they were to hearing. Thus the oldest known stapes yet discovered proved to be an almost perfect intermediate in both shape and function between the longer hyomandibular that braced the jaws and palate in fishes, and the smaller, eardrum-supportive stapes of latter tetrapods.

In 1990 the plot thickened as fossilised trackways attributed to land-walking tetrapods were found in Scotland. Other fossilised tetrapod tracks of Devonian age were then described from southwestern Ireland, although tectonic distortion had 'smeared' these prints a little, so they had to be undistorted before a meaningful interpretation could be made. These fossil trackways show that tetrapods were indeed more widespread in the Late Devonian than previously thought, and that these track makers did not overstep their hand prints, as the perpetrator of the Genoa River tracks of Victoria had done. Interestingly, these trackways showed that some of these very early tetrapods walked with digits pointing outwards from the axis of the body, not facing forwards as would be expected for a primitive fish-like limb.

Also in 1990 Clack and Coates published for first time details of the limbs in *Acanthostega* and *Ichthyostega*. What they revealed was quite remarkable, and unexpected. For neither beast had five digits, as was previously supposed. *Ichthyostega* had seven digits on the foot, whereas *Acanthostega* went one better and had eight on the hand[12]. In

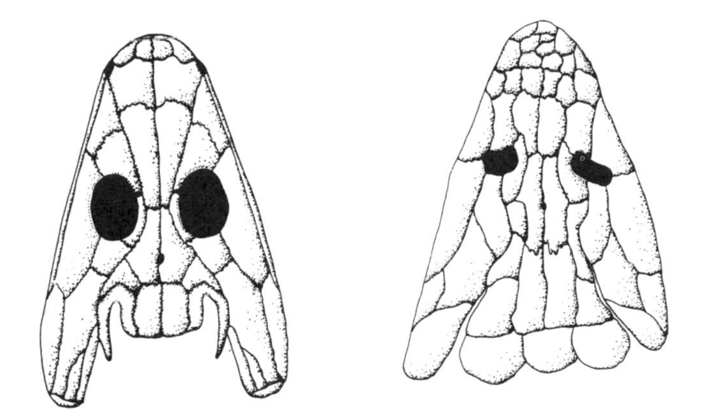

Let's play 'pick-the-fish'. Not easy is it, seeing as both skull patterns are almost exactly the same? Yet one of these has arms with fingers and legs with toes, the other has fishy fins. The skull on the left is that of the amphibian Acanthostega, *whereas the one on the right is of the fish* Panderichthys: *both are Late Devonian in age.*

all likelihood, this 'polydactylous condition' was probably more an adaptation for swimming than for walking on land. The picture was now emerging, from their studies of the stapes, the braincase, the limb skeletons and the presence of gill arches[13], that these early amphibians were really little more more than fishes with slightly modified skull patterns, and digits on the ends of their limbs. More than likely they still lived highly aquatic lifestyles, with only *Ichthyostega* possibly venturing on to land for daring, short forays into the unknown, terrestrial habitat – a land ruled by arthropods (see Chapter 5).

Following the publication of this revolutionary data and ideas on the Greenland tetrapods, the race was on to find more Devonian tetrapods, and Per Ahlberg, enthused by his participation on the highly successful East Greenland Expeditions, began looking seriously through old fossil fish collections in Britain, hunting for clues to where the best sites for tetrapods might be. He also began collaborative work with the Latvian scientist Ervins Luksevics and Russian colleague Oleg Lebedev. Lebedev had discovered the six-fingered *Tulerpeton* back in the 1980s. From this collaboration and initial searching, two major discoveries emerged. Firstly, tantalising remains of an older, more fish-like tetrapod were found amongst collections from the Scat Craig site in Scotland, previously labelled as belonging to osteolepiform fishes[14]. Secondly, a new tetrapod was identified from bits and pieces in the rich Late Famennian Ketleri Formation of Latvia[15].

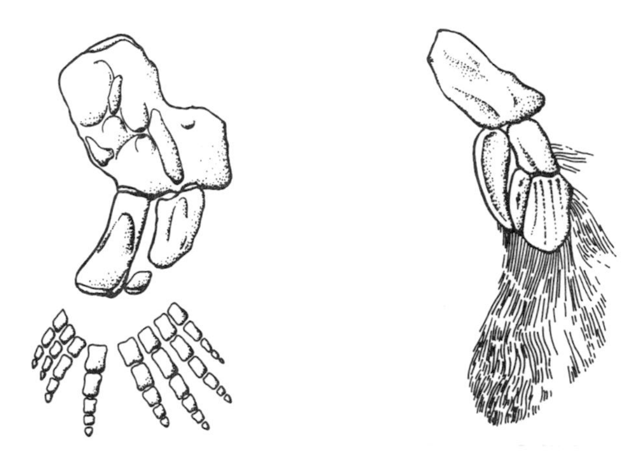

One of these is 'armless! There's not much difference between the fins of advanced lobe-finned fishes and the limbs of early amphibians. On the left is the arm and hand of the amphibian Acanthostega, *and on the right is the pectoral fin skeleton of the fish* Panderichthys. *Both have similarly shaped humerus, ulna and radius, the same bones carried on through to all land animals alive today (except those without arms!)*

In 1991 Per Ahlberg published his description of amazingly tetrapod-like scraps from the Late Frasnian (Late Devonian) Scat Craig deposit, near Elgin, Scotland. For many years these bones had been thought to have been derived from an osteolepiform fish, *Polyplocodus.* It was only through recent 'rediscovery' of the Devonian features in the new specimens of *Ichthyostega* and *Acanthostega* that it was possible for these fragments to be recognised as being of tetrapod origin. Further excavations at the site by Ahlberg later confirmed that these were indeed tetrapod remains and constituted the most primitive (or more fish-like) of the Devonian amphibians. In 1995 he named these remains *Elginerpeton*[16], and later described associated postcranial remains from the same site (most probably belonging to the same animal). *Elginerpeton* was remarkably fish-like in its anatomy, and presumably in its behaviour. Its study enabled Ahlberg also to reassess one of the Russian osteolepiforms, hitherto described as a fish called *Obruchevichthys.* Known only from a lower jaw, Ahlberg was able to recognise that this too was an early tetrapod, which he was able to place in the same primitive family which he named the Elginerpetonidae.

The Latvian amphibian was recognised from individual bones that were beautifully preserved in three dimensions, literally washed out of unconsolidated pale sands with a hose and bucket[15]. These bones were very rare indeed, and much hard work was needed to acquire sufficient bones to enable a preliminary description and reconstruction of the new beast to be made. This amphibian from the Venta River was eventually named *Ventustega curonica* and its description was published in 1994. Only the isolated lower and upper jaws, along with fragments of skull, parts of the shoulder girdle and a hip bone (ilium) were found. The lower jaw of *Ventustega* proved to be very primitive for a tetrapod in retaining large fangs on the coronoid bones, a condition otherwise seen in osteolepiform fishes, not in other tetrapods. However, work still continues at the site in the hope of revealing more of this intriguing animal.

At the same time, in 1994, the first Devonian tetrapod from North America was described by Ted Daeschler from the Academy of Natural Sciences in Philadelphia, with colleagues[17]. Based on a single shoulder girdle, *Hynerpeton bassetti* proved that by the Late Devonian tetrapods were to be found almost everywhere within equatorial–temperate latitudes.

Meanwhile, during this publication frenzy of new data on early amphibians much new work was also being contributed on the nature of the advanced lobe-finned fishes. In 1992 the fish *Panderichthys*, often thought of as the closest relative to the amphibians, was redescribed in great detail by Emilia Vorobjeva and Hans-Peter Schultze. The shoulder girdle, redescribed in 1995, was shown to be very tetrapod-like in having a long dorsal lamina on the cleithrum and reduced dorsal series of pectoral girdle bones.

In 1996 Per Ahlberg, Jenny Clack and Ervins Luksevics studied another specimen of *Panderichthys* and revealed the structure of its braincase for the first time, noting how it showed none of the specialised features usually associated with tetrapods, still retaining the intracranial joint seen in all advanced lobe-finned fishes[18]. Thus *Panderichthys* was emerging as an animal with a bizarre combination of advanced tetrapod-like features and primitive fish-like characters. Notwithstanding the striking similarity between the external skull roof patterns in *Acanthostega* and *Panderichthys*, the structure of the braincase indicated to Ahlberg and his co-workers that during the transition from fishes to early tetrapods the braincase evolved rapidly, losing its intracranial joint and undergoing modifications suited towards the development of an outer ear membrane.

In 1992 a suite of new fossil crossopterygians was described from the Middle Devonian Aztec Siltstone of Antarctica by Gavin Young,

Alex Ritchie and myself[19]. This was the sum of many years of work involving all those workers who had been to Antarctica to collect the material, then prepare and study it. Prior to 1985 almost nothing was known about the Devonian lobe-finned crossopterygian fishes of Australia and Antarctica. The importance of the new material from Antarctica was that it showed that several groups of osteolepiform fishes were present here, and that some of these represented endemic, new families. Also amongst the finds was a veritable giant of the Devonian rivers, one of the large rhizodont fishes, a monster we named *Notorhizodon* (meaning southern *Rhizodus*, alluding to its close affinity with the 6 metre long Carboniferous killer from Scotland). The Antarctic form represents possibly the earliest known occurrence of rhizodont fishes. Recently, further examination of this material by Per Ahlberg and Zerina Johanson is revealing that our original description confused the remains of two different animals. One of these is truly an early rhizodont; the other, however, based on the large incomplete skull and braincase from the top of Mt Ritchie, may well be a huge relative of *Eusthenopteron*. The Antarctic fauna also shows the earliest possible presence of primitive osteolepiforms, an endemic group called the canowindrids. This provided further indications of a possible Gondwana origin for tetrapods – bolstering the evidence furnished by *Metaxygnathus* and the Genoa River tetrapod trackways.

However, despite the evolutionary stages for the fish–tetrapod transition being better understood through detailed morphological studies, there has been little about the mechanisms that enabled amphibians to evolve from fish. One of the problems has been that for many years the apparent morphological gap between advanced lobe-finned fishes and early tetrapods was difficult to understand, because comparisons were always made between adult fishes and adult amphibians. However, study of juveniles of the osteolepiform fishes, such as those of *Eusthenopteron* that were described by Hans-Peter Schultze[20], has shown that these juveniles have more characters in common with early adult amphibians[21]. This suggests that heterochrony (changes to the rate and timing of development, in this case a reduction in the amount of development), had been a major influence in bringing about the radical morphological changes that occurred over such a relatively short geological time. Indeed, heterochrony has been shown to be a major factor in the evolutionary radiations of both fossil and recent amphibians[22]. Further ideas on the value of using heterochrony to interpret the fish–tetrapod transition have been put forward by Mike

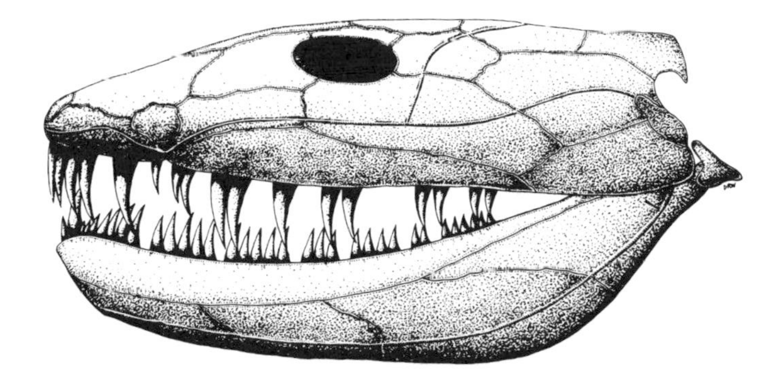

This handsome looking devil is a reconstruction of the skull of Ichthyostega, *a Late Devonian amphibian from* East Greenland *(after* Erik Jarvik*). The huge fangs of* Ichthyostega *demonstrate the highly predatory nature of these early tetrapods, undoubtedly a driving force in the evolution of amphibians that may have been the reason for them ultimately leaving the water to exploit the land.*

Coates, who in recent years has produced one of the most superb scientific descriptions of the postcranial morphology of any fossil amphibian in his work on *Acanthostega.*

In April 1996 Erik Jarvik published his monograph on the anatomy of *Ichthyostega*[23]. In that paper, which featured many illustrations of the bones of *Ichthyostega*, along with many fine reconstructions of its anatomy, are some almost disturbing pictures of humans with deformed hands and feet. These were examples of the condition known as polydactyly, in other words, these people had more than five digits on each hand or on each foot. What on Earth, you may ask, has this to do with a study of a Devonian amphibians? The answer lies in the still perplexing mystery of why did all these primitive amphibians likewise have more than five fingers and toes? Jarvik's answer was to put it down to pathogenesis – abnormal developmental malformations. He argued that as we only have a small sample of Devonian tetrapod limbs, and as they all differ, it is possible that all were diseased or deformed in some way, and that five digits was the normal condition even then. This answer is, to say the least, not wholly satisfying to most researchers in the field, who consider polydactyly in amphibian limbs to be the normal condition. Mike Coates has highlighted the fact that in all known Devonian tetrapods, the hand and feet were polydactylous, not just in one or two specimens[24]. In *Acanthostega*, for example, the hand and foot both have eight digits. *Ichthyostega*, on the other hand, had seven, but more importantly had the same conserved pattern of four stout and three slender digits. This shows it to be very unlikely that the extra digits were pathological abnormalities.

Developmental biologists have approached the problem from a different angle, explaining such anomalous patterns as perfectly normal developmental events involving a major evolutionary transition arising from critical genetic variation. For example, Neil Shubin of the University of Pennsylvania and Per Alberch of the Natural History Museum in Madrid have pointed out that the limbs of embryos of living primitive tetrapods resemble the bifurcated fin arrangement in fishes[25]. The main axis of the fish fin has to then turn 90° during the evolution of the tetrapod limb to enable the animal to bear the body weight and walk on land. During development the last digits to form on the tetrapod limb are the most distal (anterior) ones. The loss of some of these during the stabilisation of the tetrapod hand and foot pattern to five digits came about through the failure of condensation of some digits very early in development. Regulation of developmental processes is strongly influenced by Hox genes, which determine the pattern of structures.

The expression of some of these genes has been shown to be related to the major skeletal boundaries in the hand, and malformations resulting from experiments on these genes relate principally to the timing of expression of particular skeletal elements, rather than their position in the hand or foot. A recent review by Shubin and colleagues has outlined the possible roles that Hox genes have played in the development of invertebrate appendages and vertebrate limbs[25]. Once again we see the all-pervading theme of heterochrony, tinkering with the timing of development, and directing the construction of new morphologies, and ultimately new species. Developmental studies of Hox gene expression in the limbs of tetrapods and fishes have revealed that the digits of tetrapods are novel structures. In other words, they are new structures that are not related in any way to preexisting structures in the advanced lobe-finned fishes. Digits are not the homologues of fin rays.

When they studied the embryological development of the fin of the teleost zebrafish *Danio rerio*, Paulo Sordino, Frank van der Hoeven and Denis Duboule of the University of Geneva found that the first condensation of cells of the fin bud occurs as a thickening and growth of patches of mesodermal cells[26]. These are enveloped in an ectoderm layer, much like in tetrapod limb buds. As development of the limb proceeds, the ectoderm very rapidly protrudes and then folds in on itself. Dermal skeleton then appears inside the fold. As it does so there is a rapid reduction in mesenchymal cell production. When this transition takes place dictates how much endoderm forms, compared with ectodermal skeleton. Fin rays are formed of ectodermal skeleton, whereas digits are endodermal skeleton.

This transition occurs very quickly in the pelvic fin of the zebrafish, so that there is very little, if any, endoskeleton formed. Hence teleost fishes sprout fins, not arms and digits. In tetrapods, however, this change from endoskeleton to exoskeleton production fails to occur, and the limb is composed entirely of endoskeleton in the form of humerus, ulna, radius, and so on. Each limb bone forms in sequence, with the final elements being digits. These are the terminal expression of the endoskeleton limb. Lobe-finned fishes, like the Late Devonian *Eusthenopteron*, show an intermediate condition. During their development there was a delay in the onset of the folding of the ectoderm, resulting in a longer period for more endoskeleton production. The result is the formation of a humerus, radius and ulna, as well as a femur, tibia and fibula. However, because the onset of ectoderm production is still turned on, albeit later than in teleost fishes, fin rays grow from the endodermal bones.

One mystery associated with the fish–tetrapod transition has been the place of origin for the first tetrapods. Although earlier I had advocated a possible Gondwanan origin, based on the Genoa River trackways, the *Metaxygnathus* jaw and the primitive, early appearance of some osteolepiform fishes[21], the argument that panderichthyids gave rise to tetrapods presents an anomaly. The several forms of panderichthyid fishes are all from the Frasnian (Late Devonian) of the Euramerican region, which at that time, 370 million years ago, formed a single continent, quite distinct from Gondwana. This comprised what is now western Europe, Baltica, North America and east Greenland. Along with the presence of elginerpetontids in the Late Frasnian of Scotland and Russia, plus the abundance of slightly younger tetrapods in Latvia, Russia, east Greenland and North America, the arguments for a northern origin for the tetrapods seems strong.

It now seems more than likely that the fish–tetrapod transition occurred in this northern region, and the early tetrapods then became widespread by the Famennian (latest Devonian), spreading southwards into Gondwanan regions[27]. This implies the marine barriers were probably not a hindrance to their migration. The six-fingered amphibian *Tulerpeton* was found in marine limestones of the Russian Platform, and although this is not directly indicative of where it lived when alive, it doesn't seem inconsistent with the fact that the fish *Panderichthys* inhabited shallow marine environments. So, why not also its descendent tetrapods?

The fish–tetrapod transition has often been seen as a very complex transition involving massive degrees of morphological change, such as the ability to breathe from air, rather than gill breathing in water; the ability to walk on land, rather than swimming; the ability to hear as opposed to sensing with a lateral-line system; the ability to retain body moisture rather than having to exist under water; and the ability to reproduce out of water, amongst other things. If these steps were taken one at a time, it is not so difficult to understand that the first tetrapods, did not rush out of the water, but emerged fully equipped for a life on land, already preadapted to this new habitat.

The very tetrapod-like panderichthyid fishes are thought to have been predators, possessing very large heads relative to their body size, large fangs, and the ability to open their mouths wide. In their ancestors this propensity for predation resulted in the evolution of a hinged braincase[27]. This feature characterises the 'crossopterygian' group, but it was later to close up in several independent lineages – one being the lungfishes, the other the tetrapods. Perhaps it was these predatory adaptations in the panderichthyid fishes that drove them toward a tetrapod-like morphology. The stronger hands and feet, if not initially an adaptation for walking on land, could well have been adaptations for improving speed whilst swimming. They would have allowed the beast to leave the water for short periods to escape predation pressures in the water and breed somewhere nearby in relative peace. The eyes on the top of the skull would have enhanced their ability to view the new world, terrestrial land, and watching for passing tasty morsels that ventured near the water's edge.

The long fish-like tail of the first tetrapods was practically unchanged from that in panderichthyid fishes, as it was these animals' main propulsion for many million of years to come. Even after the more advanced amphibians developed better limb girdles and could walk freely on land, the tail was still the important propulsive device for their aquatic forays. Finally the scaly cover of fishes was not lost on the first amphibians either. As they began to invade the terrestrial habitat, equipped only with poor limbs capable of limited terrestrial locomotory powers, the scale-cover over the belly would have been an important protection for the animals' viscera, when dragging themselves across the ground.

So, one of the great steps in evolution, the invasion of land, probably did not come about until well after the first tetrapods had fully evolved from fishes. The refinement of their limb skeletons and vertebral columns took place during the Carboniferous Period, the age when the first reptiles evolved. Such creatures acquired a great evolutionary novelty, the hard-shelled amniote egg. Tetrapods were then no longer tied to water. They were able to venture for the first time away from the rivers and lakes of the humid Carboniferous jungles, and invade inland habitats.

The gap in the fossil record of the first Early Carboniferous tetrapods has been filled in the last few years with the discovery of new amphibian faunas in North America, Britain and, most recently, Australia[28]. So, as the fish–tetrapod transition becomes much better understood, not only from a wealth of new fossil evidence, but also from advances in developmental biology, the transition from fish-like aquatic tetrapods to fully terrestrial forms is now coming under close scrutiny. The huge radiation of tetrapods during the early part of the Carboniferous is as much a mystery today as the fish–tetrapod transition was about 10 years ago. Hints of new discoveries are emerging from laboratories in Australia, Britain and the USA. Watch these places. I predict that if work progresses as fast as the Devonian tetrapod research, then the mystery of amphibian radiations and the origins of true terrestriality in tetrapods will be soon be resolved. Once the dogs are pointed to their prey, there is no stopping them . . .

9

Say it with Flowers

Evolution in plants

I'm sure that you are familiar with Murphy's Law – the one that says that if anything can go wrong it will go wrong. But that's only one of Murphy's many laws. In most fields of endeavour there are numerous subsets of Murphy's laws. Whether they have ever all been codified, I doubt. In the field of palaeontology there are quite a few, as there are in publishing. One of the more common ones in publishing relates to the fact that you have read through squillions of drafts of your manuscript, it's been read and checked by referees, then by editors, and you've exhaustively checked the page proofs, until you are sure it is the most perfect manuscript ever to grace a printing press. Then comes the day of publication. I had one of those days today. The second edition of my book on stromatolites was published. I opened the pages tentatively, waiting for the devilish hand of Murphy to strike. And strike he did. Within less than a minute I had found it – a spelling mistake. It shouts at you from the page and seems to be printed in a font 20 times larger than all the other words, so you are sure that is the one word that everyone will see. But Murphy didn't always get it right. There is one of his laws which he might think will keep his reputation as a spoiler of life flying high. But it has backfired on him on a number of occasions. It all relates to the finding of the best fossil specimen. Let me tell you of such an example, when Murphy was hoist by his own petard.

It was late in the afternoon. We had been driving south all day to the Kennedy Range region of Western Australia on washed-out tracks.

On reaching Mt Sandiman station we headed west towards the scarp of the Range. The tracks had virtually disappeared. We were chasing some outcrops of a 40 million-year-old sandstone in which a mixture of marine fossils and plant fossils had been found. What was special about this ancient beach deposit was that amongst the plant fossils that had been collected a few decades earlier was a rather badly preserved fossil fruit of a banksia. A typical member of the Proteacea, that great group of Southern Hemisphere plants that includes grevilleas and proteas, banksias produce a very hard, woody cone-like fruit that bears seed-bearing follicles. The one specimen that had been found in these deposits represented the earliest example of a banksia, as with just the leaves and pollen you would be hard pushed to know whether you had a banksia or not. What we hoped to find was a decent specimen that we could describe and which could provide an indication of the nature of some of the earliest of these proteaceous plants.

It had been one of those grey, overcast, depressing sort of days. But as we finally reached the scarp just before sunset, a tiny break in the cloud appeared on the western horizon, and the red Permian rocks that underlay the younger beach deposits were suddenly set on fire against a backdrop of a purple–grey sky. We had two days to find the site from where the fossil banksia had come. Surely that would be enough time. The first day we trekked for miles along the scarp. Lots of good molluscan fossils, and some excellent tiny sea-urchins. But no plants. The next day we ventured further afield, out on to the plains in front of the scarp. But still no plants. Lunchtime came and went, and the day began to sink away. Evening came and we still hadn't found any.

We decided to give it one last half an hour at the beginning of the third day before we began the long drive south to Perth. There was one small outlier that we hadn't checked. It looked no different from every other scree-covered hill we had climbed. It took us a quarter of an hour to get there. Then Murphy struck. He knew we had just 15 minutes left before we had to go. The scree was covered in large, incredibly hard, silicified sandstone boulders. And grinning up at me from almost the very first boulder I looked at was the most perfect, huge, wonderfully preserved banksia 'cone' you could imagine. The actual 'cone' (technically an infructescence) was no longer there but it had left a faithful impression of itself in the sandstone. But it was time to go. Naturally the fossil was lying in the middle of this large 40 kg boulder. There was nothing for it but to carry it the kilometre back to the vehicle. That would be one in the eye for Murphy.

This is Banksia archaeocarpa, *drawn at actual size, the oldest fossil fruit from a banksia, and, indeed, the oldest species of the genus so far discovered. Its fossil remains were found by one of us* (Ken) *through serendipitous good luck, searching on the last day of a three-day expedition.*

How I managed it and was still able to walk upright ever again I'm not sure, but when you are carrying the palaeontological equivalent to what at the time seemed to be the Crown Jewels, it's amazing what the tortured human frame can manage. Once a latex rubber cast had been made of this natural mould back in Perth it became even more apparent what an amazing fossil this was. Not only was it beautifully preserved, but it was remarkably similar to banksia 'cones' from trees still living in Western Australia today. My botanical colleagues at the local herbarium were very surprised by this similarity. They, like most biologists, had the attitude that an early member of a lineage must be simpler than later ones. Evolution, it is usually thought, can only proceed from the simple to the complex. But this is certainly not always the case, and the banksia fossil showed this in a most elegant fashion.

The banksia inflorescence is actually made up of many separate flowers with lots of individuals making up the whole structure, which looks very much like a flaming candle. Living banksias have flowers that are often bright yellow or orange in colour. What colour was our 40 million-year-old flower? There's no way of knowing. Maybe, like its close look-alike modern-day cousin, *Banksia attenuata*, it was bright yellow. The more individual flowers, the more bracts on the infructescence, then potentially the more seeds that can be formed. This ancient fossil was like many living forms in being large, complex and with lots of bracts and seed-bearing follicles. Nothing at all like the simple form that many botanists would have expected from a specimen of such antiquity. But evolution can equally well go in the direction of decreasing complexity as well as to increasing complexity. In other instances, such as this banksia, there may be very little change over 40 million years. My colleague John Scott and I decided, however, that this fossil banksia was sufficiently different from any living species to warrant its very own name. We called it, not surprisingly, *Banksia archaeocarpa*, meaning 'old fruit'[1]. This is the oldest unequivocal banksia known to date.

Forty million years, and little change. Quite impressive, you might think. But another find, made more recently in Australia, makes this pale into insignificance. Whereas we have contributed to our understanding of the nature of the evolution of one particular group by a fossil find, this more recent botanical discovery worked the other way – the discovery of a tree, still living and just a few hours' drive from Sydney, the nation's largest city. A tree that had lain undiscovered, and without an intervening fossil record, for over 50 million years.

Working his way deep through Wollemi National Park in the Blue Mountains in 1994, less than 200 km northwest of Sydney, David Noble, a New South Wales National Parks ranger, came to a sheer 600 metre drop. Not one to pass up such a challenge, he abseiled, with two colleagues, into the damp rainforest far below. If a lost world exists anywhere on Earth, it must have been here, for emerging out of the understorey of ferns were some extremely strange, giant trees, the likes of which Noble had never seen before. These trees were huge, up to 40 metres high, and with a 3 metre girth. The leaves were waxy and very fern-like, and the bark of the tree had a peculiar character, as though festooned with a covering of chocolate bubbles.

Once a full set of leaves, cones and bark had been collected and studied by botanists in Sydney, it soon became apparent that here was an entirely undiscovered, and exceedingly rare, type of araucariacean pine. Known as the Wollemi pine, it is represented by only a single population of about 20 mature and a similar number of immature, trees. It was officially named *Wollemia nobilis*, in honour of its discoverer[2]. What was even more remarkable was that the only form that was even remotely similar was a fossil of a plant known to have lived in Tasmania and New Zealand more than 50 million years earlier. Before the angiosperms (the flowering plants) evolved, the world was dominated by trees such as araucariacean pines – it was a world coloured green. If the incredibly close morphological similarity between the leaves of this tree with the fossil leaves indicates a true evolutionary relationship, then it would indicate that this is a true relict from the time when dinosaurs roamed the Earth. A time before flowers existed.

Amazingly at almost the same time that the Wollemi Pine was discovered, a similar discovery was made in Queensland. In 1961 Bernie Hyland, of the CSIRO, Australia's national scientific research organisation, collected a number of nuts from a rainforest on the slopes of Mt Bartle Frere in northern Queensland. For more than 20 years Hyland puzzled over these nuts, that were unlike anything else that had been described. Eventually Hyland's colleague Bruce Gray managed to work out that they came from a macadamia-look-alike rainforest tree. But what on Earth was it? Then in early 1995 Andrew Douglas of the National Herbarium of Victoria in Melbourne brought some of these specimens back with him to study. Because of their enigmatic nature he showed them to palaeobotanist Andrew Rozefelds who immediately recognised them. But not from any living plant. He had seen the selfsame nuts illustrated in fossil form in an article published in 1875, describing some fossils, possibly Early Miocene in age (about 20 million years) from near Ballarat in Victoria, at the other end of the country from northern Queensland.

The tree, a member of the Proteaceae, like the banksia, grows to about 35 metres tall, but belongs within its own subfamily. Like the Wollemi Pine, the Mt Bartle Frere tree is extremely rare, being known from just five groups of trees, each of which have just four or five individuals. The lack of any juvenile trees in these stands points to this incredibly rare species being on the verge of extinction. One reason for this may be that of the hundreds of seeds that have been collected, all have a pair of holes in them, thought to have been made by the giant white-tailed rat, *Uromys caudimaculatus*. Interestingly, the fossil nuts show no such sign of predation, suggesting that the tree evolved in a giant white-tailed rat-free environment. Another reason may lie in the fact that this nut is incredibly hard, making it difficult to germinate. It is possible that for it to do so it has to pass through the gut of some appropriate animal. Perhaps that animal has become extinct, and the tree may well follow suit.

To unravel the evolutionary history of any group of organisms it is always important to have the oldest known fossil from any particular group. For that reason the scientific literature is often festooned with articles proclaiming to have found 'the earliest record of a lesser-spotted cordwangler' or some such beast. Plants are no exception, and in the hunt for the oldest, and therefore the one most likely to shed most light on the origins of the group, much effort has gone into discovering the oldest angiosperm, or flowering plant. The quest for finding the oldest flowering plant is almost like a quest for the palaeontological Holy Grail. It's not just that we want to understand when a particular major group of plants first evolved, a group that, after all, gave rise to daffodils, daisies and dahlias; or to be able to unravel the evolutionary history of the group and be able to work out, perhaps, which earlier group of plants it evolved from, and the evolutionary mechanisms that allowed its evolution. No, it's not just all those things that make the hunt for the earliest flowering plant something special. It is the fact that what the evolution of the earliest flowering plant really marks is a major breakthrough in evolution of the landscape – the evolution of a wide spectrum of colour.

Since the first vegetation began to cover the land more than 400 million years ago, the Earth rapidly became a green place. And green dominated the world for more than 250 million years – until the first flower evolved. What colour it was, we don't know. Maybe it was just white, but in all likelihood, once the breakthrough had been made, and the explosive radiation of flowering plants took place during the latter part of the Cretaceous Period, a whole range of flower colours probably evolved. In the early twentieth century, ideas about the nature of the first flowers centred on large floral structures, like those possessed by the living *Magnolia*. However, recent finds from the fossil record indicate that the earliest flowers were probably much smaller. For instance, the most extensive deposits in the Early Cretaceous (120 million years old) of Portugal and eastern North America show that the flowers were generally less than 2 mm long[3]. Interestingly, the earliest unequivocal evidence for flowering plants comes from magnoliid-like pollen found in

125 million-year-old rocks in Israel and southern England. By about 90 million years ago a wide diversity of magnoliids, rosiids and hamamelidids (plane trees and the like) had evolved. Monocots, like ginger and palms, had also evolved by this time.

The last few years has seen the discovery of a number of new sites, that have yielded new information on the early evolution of flowering plants. For instance, in 1993 scientists at the Nanjing Institute of Geology and Palaeontology in China described five different species of flowering plants, of Early Cretaceous age (that's about 120 to 125 million years old) from Heilongjiang Province[4]. These predate a number of other records from rocks about 110 million years old in Russia and the USA, making the Chinese fossils amongst the oldest records of flowering plants so far known. The Chinese fossils are preserved as entire leaves and were derived from plants that appear to have been inhabiting warm, humid swamps in association with ferns, conifers, cycads and ginkgos. Flowers, however, have not been found. It was in such a warm, stable environment that flowering plants are first thought to have evolved.

Although leaves, pollen and stems are the parts of plants most usually fossilised, as I have shown, hard, woody fruits can also be fossilised, given the right conditions. Occasionally, even flimsy, delicate flowers can be locked into stone forever. There are currently two candidates for the oldest flowers, one from England, and one from Australia. Indeed, it was the discovery of this fossil flower in Australia that spurred English palaeobotanist Chris Hill to search for, and then find, what could be the world's oldest flower.

The plant, called *Bevhalstia pebja* (named after the late English palaeontologist Bev Halstead), occurs quite commonly as impressions of complete plants in a fine clay. It is thought by Hill to have been an aquatic form up to 25 cm long[5]. The leaves were short and seedling-like. A number of small, flower-like structures have been found, each about 7 mm across. However, proving that they really were flowers is another matter. Certainly their structure and position on the plant is very suggestive that they were indeed flowers. Support for this interpretation comes from slightly younger flower-like structures found in the Aptian (115 million-year-old) deposits at Koonwarra in Victoria. David Taylor and Leo Hickey, of Yale University, have interpreted these structures, which look very similar to the English fossils, as flowers. They consider that the ancestral angiosperm would have been a small, rhizomatous perennial, with tiny reproductive organs[6].

The very simple 'flowers' on these plants are not adorned by petals or sepals, being similar in some respects to those in a group of cycad-like plants called Bennettitales. It is generally accepted that the angiosperms are most closely related to the Bennettitales, and an obscure group of living plants called Gnetales. The three groups comprise the 'anthophytes'. Hill believes that *Bevhalstia* was an aquatic plant. If so, this has important implications for the origins of flowering plants as a whole. An aquatic origin would perhaps explain why angiosperm pollen has not turned up very often in deposits of this age.

The question of exactly how flowers evolved is still essentially unanswered. Likewise, the origins of elements of the flower, the stamens and carpels, is also problematic. Rather than being complex structures, angiosperm stamens are more simple than pollen structures in Bennettitales, Gnetales and Mesozoic ferns. The carpels, on the other hand, are relatively more complex[3]. But don't be fooled into thinking, however, that it wasn't until flowers evolved that plants struck up their very intimate relationship with insects. Far from it. While insects may have rapidly evolved to become plants' willing sex slaves, many groups of insects had enjoyed cosy relationships with plants for a very long time before. There seems little doubt that the apparently rapidly established role of insects as angiosperm pollinators led, in part, to the rapid diversification of the group early in its evolutionary history. This is shown by the almost simultaneous radiation of both angiosperms and pollinating insects. Recent studies have pointed to the possibility that this quick link-up between pollinating insects and angiosperms arose because these insect groups had already become established as plant pollinators well *before* flowering plants had evolved. Insects have been discovered playing a significant role in the pollination of some living cycads[7], suggesting that they may have played a similar part in pollinating some of the Mesozoic cycads. Moreover, it has been suggested that the extinct, cycad-like Bennettitales had to have been pollinated by insects because of the nature of their reproductive organs. Such relationships may go back as far as the Triassic Period. Certainly, the nature of some bennettitalean fossils in the Late Jurassic point to their pollination by insects at this time[7].

Insects that are involved in plant pollination, such as bees, butterflies and moths, beetles and flies, show patterns of evolutionary diversification that indicate their coevolution with angiosperms. The bee *Trigona*, for example, has been found in 96–74 million-year-old strata, suggesting that bees were present during the critical radiation of the angiosperms in the Early Cretaceous[7]. The Lepidoptera (butterflies and moths) are known to range back to the Early Jurassic, about 190 million years ago, well before the first known angiosperm. Their evolutionary radiation, however, corresponds well with that of the angiosperms, in the Early Cretaceous. The appearance of the Lepidoptera before the angiosperms supports the view that the long proboscis, that plays such an important role in flower pollination, was originally adapted for the uptake of water or resin, rather than nectar.

The fossil record also reveals more direct associations between plants and insects, other than that involving pollination. Leaf damage preserved on 97 million-year-old leaves from the mid-Cretaceous Dakota Flora, described by Conrad Labandeira and colleagues from the Smithsonian Institution, indicates the activity of leaf miners much earlier than had hitherto been recorded[8]. Leaf miners are the larvae of insects that spend their preadult lives eating out and inhabiting tunnels within the leaf. Different species leave different, very distinctive traces of their meanderings through the leaf as they grow. Thus within just 25 million years of their evolution, angiosperm organs were being exploited in a complex way by contemporary insects, in much the same way that they are today. However, such examples are rare. The general absence of indication during the early period of angiosperm radiation of insect herbivory on these plants has been attributed in the past to the lack of an adequate fossil record of plants of this age. However, the rich Cenomanian Dakota Flora is so well preserved that the identity of the insects that perpetrated the damage can be made to the genus level, revealing that they are attributable to living taxa. Moreover, the floral structures of the plants indicate that a variety of pollinators were active at this time.

The dominant angiosperms that comprise the Dakota Flora are magnolias, laurels and hollies. Leaf miners, identified in the leaves by Labandeira and colleagues, are the distinctive *Stigmella* and *Ectoedemia*. The impressions left by the activity of these insects is so good it is possible to see the oviposition site, the pattern of faecal pellet distribution, the enlargement of the mine as the leaf miner moulted into larger and larger sizes, and the size, shape and internal structure of the larval chamber. These leaf miners are the larvae of a family of moths, the Nepticulidae. The activity of many of these leaf miners is incredibly similar to the activity of their living descendants. Perhaps of importance to assessing the role of pollinators on these early flowering plants is the fact that the adults of these leaf miners today have mouth parts which are used for nectar feeding. If these early ancestors fed likewise, then they are likely to have also played a role in pollinating the plants that, like them, their juveniles fed from.

Yet, the activity of leaf miners was not coincident with the rise of the angiosperms. Leaves of the voltziacean conifer *Heidiphyllum* from the Late Triassic in Queensland, Australia, have been found with evidence of leaf miner activity by Andrew Rozefelds and Ian Sobbe of the Queensland Museum[9]. These represent the earliest evidence of leaf mining known. Other examples of leaf mines have been recorded in the seed-fern *Pachypteris* from near the Jurassic/Cretaceous boundary in north Queensland, possibly made by nepticulid moths. With the diversification of flowering plants in the Cretaceous, an ever-increasing larder became available to groups of Lepidoptera whose larvae were adapted to feeding on leaves.

To find the earliest evidence of insect herbivory we have to travel much further back in time. Conrad Labandeira and T. Phillips of the University of Illinois have described the fossils of seven galled tree-fern petioles from 300 million-year-old floras in Illinois[10]. These galls are swollen abnormalities that occur in the basal petioles that support the fronds of the marattialean tree-fern *Psaronius* and were probably made by the larvae of some sort of holometabolous insect (these are insects that have a pupal stage between the larval and adult stages, resulting in two metamorphoses taking place during development). The trouble is that the fossil record of all these insects, the scorpion flies, flies, moths, butterflies, lacewings, wasps, bees and ants, indicate earliest occurrences no older than the Early Permian. It therefore seems likely that at least one of these groups was present in the Carboniferous. In addition to galls, plants of Late Carboniferous age also show signs of herbivory from the activity of some insects piercing-and-sucking internal vascular tissue, as well as external damage to fronds. From such evidence Labandeira and Phillips believe that tree-fern forests of this time had well-established, diverse insect dietary guilds. Indeed, they point out that, apart from the absence of leaf mining, there appear to have been modern-style insect communities living in and on these tree-ferns 300 million years ago.

But is there any earlier evidence? Strangely, there is little direct evidence for insect herbivory in the fossil record from when vascular plants first evolved in the Late Silurian about 410 million years ago, until well into the Carboniferous Period, about 330 million years ago. The only possible evidence may be damage to stems of *Rhynia* from the

Rhynie Chert, that may have been made by arthropods. William Shear of Hampden-Sydney College, Virginia, has suggested that one of the reasons for the paucity of evidence of herbivory might be that by-products from the synthesis of lignin, which was present in early vascular plants, were toxic to early terrestrial animals. So true herbivory could only have evolved when animals developed enzymes and a gut microflora of symbiotic bacteria which could bypass the need for an intermediary decomposer and break up fresh plant material more directly[11].

However, recent fossil evidence suggests that during this long period of apparent lack of exploitation of plants, parts other than leaves were the targets of peckish little vegetarian invertebrates. The new evidence is indirect, coming in the form of coprolites (fossil faeces) discovered in Early Devonian (390 million years) and Late Silurian (412 million years) rocks from the Welsh Borderland. A team led by Dianne Edwards of the University of Wales and Paul Selden of the University of Manchester discovered fossilised undigested land-plant spores[12]. This led them to suggest that spore eaters may have existed at these times. Others have suggested that spores and pollen may have provided a food source before other parts of plants were exploited. This is not an unreasonable suggestion, given the high energy value of such food. The big question, however, is whether the animals were feeding directly from the plant, or were they just detritivores, such as millipedes, that had been feeding on litter rich in spores. The lack of degradation of the spores points to primary feeders. But just which animals were the creators of these coprolites is the 390 million dollar question. These deposits predate the earliest known insects, so we have to invoke arthropods other than insects, or perhaps suggest that insects had evolved even earlier than we had hitherto thought.

The earliest undoubted vascular plants were lycophytes such as *Baragwanathia* and the rhyniophytes *Salopella* and *Hedeia* from Victoria in Australia, lycophytes and *Psilophyton* from Libya, and *Cooksonia* – perhaps the most primitive-looking of all plants – all from the Late Silurian (around 410 million years ago). Some doubts about the terrestrial status of Silurian forms of *Cooksonia*, long thought to be the earliest land plant but lacking vital evidence, were dispelled by Dianne Edwards and co-workers from the University of Wales. They found water-conducting vessels (tracheids) and stomata in *Cooksonia* from Late Silurian rocks in the Welsh Borderland, sure evidence that the plants would have been capable of supporting their weight on land.

Baragwanathia and *Cooksonia* are usually preserved in marine sediments, suggesting that these plants grew near shore. Envisage a time before this, when the only plants on Earth would have been a few patches of mosses and liverworts, occupying damp nooks and crannies. The land was essentially rock-coloured: greys, blacks and browns. But as vascular plants took hold near the water's edge a green mantle soon spread across the world. That this was a fairly rapid event is shown by the presence of vascular plants in deposits from inland sites, such as the 395 million-year-old hot-spring Rhynie Chert deposit (see Chapter 5).

But the spread of plants across the world, and to such upland sites, is thought to have had a tremendous impact not only on the nature of the landscape, but also on the atmosphere. Greg Retallack of the University of Oregon has described the world's oldest well-differentiated forest soil, perhaps a little ironically, from a continent now lacking in any trees – Antarctica[13]. In the 380 million-year-old Aztec Siltstone in Victoria Land, Retallack found a red, clayey palaeosol, complete with root traces up to 11 mm in diameter and 1.5 metres deep. These formed in what are thought to have been a subhumid seasonal climate. Similar soils today are found in the Indo-Gangetic plain of northern India. Combined with evidence from earlier soils, there is mounting evidence that the spread of plants to upland sites was accompanied by an overall increase in size of the plants as they diversified, in particular a lengthening of the roots. This can be empirically demonstrated from the fossilised remains of root passages in these ancient fossil soils. From the Silurian to the Devonian they become longer and increase in width and density.

The effect of this would have been an acceleration in the chemical weathering of rocks, the consequence being an increased removal of carbon dioxide (CO_2) from the atmosphere. This is because the effect of silicate weathering is to convert soil carbon, a product of photosynthesis, into dissolvable bicarbonate. This is then carried to the oceans by rivers and precipitated out as carbonate in the sediments, ultimately being trapped in limestones. The decline in atmospheric CO_2 has been empirically demonstrated from the isotopic compositions of carbonates in the red palaeosols. Retallack considers that such palaeosol evidence is suggestive of 'widespread carbon-hungry ecosystems in well-drained soils by Silurian times'. Independent confirmation of these changes has comes from another source: the number of stomata in fossil leaves. Because plants open and close their stomata in order to regulate their CO_2 uptake, Jenny McElwain at Royal Holloway College, London,

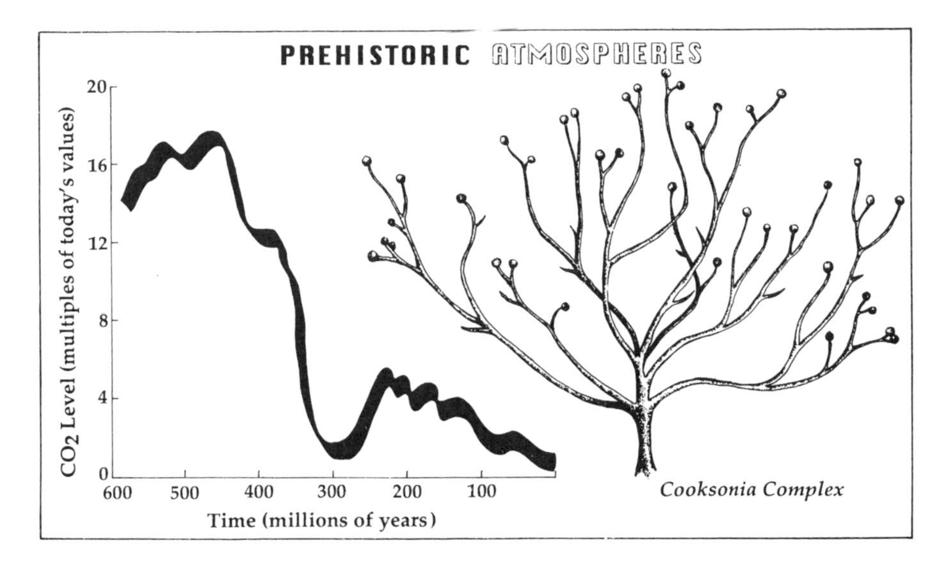

Beware the greenhouse effect, you say! If you think a few cows breaking wind and few smoky industrial chimneys are causing havoc with the climate today, just take a look at past CO_2 levels! Back about 450–500 million years ago, CO_2 levels were more than 16 times higher than today, so the greenhouse effect would have been very severe. Such atmospheric conditions made the first plants, like the little Cooksonia, *very happy to leave the seas and colonise the land.* Cooksonia *had stomata, proving that it was a subaerial plant and not living in shallow waters.*

with her colleague, Bill Chaloner, found that the higher the atmospheric CO_2 levels, the fewer the stomata[14].

The consequence of removing so much CO_2 from the atmosphere would have been the reduction in global greenhouse effect, combined with lowered global temperatures. This in turn would have caused less runoff into rivers, which would have had a decelerating effect on global weathering, so helping to balance the acceleration in weathering activity caused by the diversification of the plants. It has been estimated that the tremendously large drop in CO_2 levels in the atmosphere between 400 and 360 million years ago was from nearly 10 times present-day levels down to levels similar to today's. This dramatic decline has been attributed almost entirely to the spread of vascular plants, but in all likelihood it was also due to increases in plant size. The ultimate effect of plants greening the world and the resultant great drop in CO_2 levels, was the initiation of a major global glaciation in Permo-Carboniferous times.

Despite modern fears of increased global warming through the so-called 'greenhouse effect', caused by the geologically instantaneous appearance of humans' industrial activities, many aspects of the fossil record show that not only have global temperatures fluctuated dramatically over millions of years but so too have CO_2 levels. We are probably still coming out of the last great Ice Age; so is it really any surprise that global temperatures should be showing an upward trend? Certainly, separating the effect of naturally changing climatic cycles from perturbations that might have been caused by human influence is going to be a major challenge facing scientists as we march into the twenty-first century.

10

Monsters of the Deep

New advances in the evolution of Mesozoic marine reptiles

In the quiet waters of Hamelin Pool at the southern end of Shark Bay in Western Australia lurk stromatolites. These brooding mounds of carbonate offer a wonderful window into very early life on Earth. The crystal-clear waters that lap gently round and over these structures are very inviting on a hot Australian day. Never mind that the sea here is twice as salty as normal seawater, it's still a great place to go snorkelling. And when you dip below the surface you really are transferred back a few billion years. Well, that is if you ignore the one species of fish that seems to be able to cope with this highly saline water. But if you are really lucky, or really unlucky, depending on your point of view, you could encounter other animals that have been swimming in the seas for nearly a hundred million years, and which have seen their relatives succumb to the finality of extinction – sea snakes.

They have reputations for being curious beasts. They also have a reputation for hanging out around the stromatolites and for being extremely venomous. So, even though when one screams out of the deep blue straight at you, 2 metres of deadly grace and elegance, you generally don't take time to ponder on why sea snakes are the only fully marine reptiles left in existence. You set out rather rapidly, to prove that humans can, indeed, walk on water.

Although other reptiles live in the sea, like crocodiles and turtles, neither of them have broken the metaphorical umbilical cord that ties them to the sea. But sea snakes have. And as we shall discuss later in this chapter, recent research has thrown up a surprising reason for why sea snakes seem to be so at home in the marine environment. But what of their distant reptilian relatives – the huge monsters of the deep, the plesiosaurs, ichthyosaurs and mosasaurs, that cruised the oceans along with snakes in the Cretaceous? They are now long extinct, all having gone by the the end of the Mesozoic Era. Yet their fossilised remains still emerge from aquatic tombs now stranded on land, and are yielding much information on the relationships and lifestyles of these monsters of the deep.

Plesiosaurs have long been known as fossils from England, when the Reverend William Daniel Conybeare (1787–1857), then living in Bristol, made a study with his colleague Henry de la Beche (1796–1855) of the Liassic ichthyosaurs and plesiosaurs of southwestern England[1]. It was in 1821 that they proposed the name *Plesiosaurus.* From this the general term for the group, the plesiosaurs, has been derived. The famous fossil hunter Mary Anning found England's first complete plesiosaur in 1823 in the 200 million-year-old Early Liassic of Lyme Regis, in Dorset. This superb specimen is now housed in the Natural History Museum in London. Despite being one of the first plesiosaurs ever found, it was only in 1997 that the specimen was formally described and a correct diagnosis for the species given[2]. Although many species of the genus *Plesiosaurus* have been described over the years, only the original species, *Plesiosaurus dolichodeirus*, and maybe one or two other species, can now be truly called *Plesiosaurus.*

The plesiosaurs, along with the dolphin-like ichthyosaurs, are perhaps the best known of the marine reptiles that lived in the Mesozoic Era when dinosaurs trod the land. However, amongst the marine reptiles there is another group that concerns us here – the mosasaurs. These were large, lizard-like creatures with paddles and long tails that had a huge head armed with many sharp teeth. The mosasaurs have long been thought to be marine lizards, being close relatives of the modern varanid lizards, such as the Australian goannas and the Komodo Dragon of Indonesia. Not so any more. Recent research is shedding much new light on the origins and relationships of some of these groups, much being based on incredible new finds of marine reptiles of truly gigantic proportions.

Ask anyone what was the largest animal that ever lived, on land or sea, and the most common, and correct, answer is the blue whale, *Balaenoptera musculus.* This creature reaches lengths of nearly 31 metres with an estimated body weight of around 130 tonnes. The largest of the underwater predators is the sperm whale, of Moby Dick fame, that reached a total length of 18 metres. Until recently the prehistoric marine reptiles were no match for the modern whales in terms of their size. The largest of the predatory short-necked plesiosaurs (called pliosaurids) was either *Kronosaurus* from Queensland in Australia, or *Liopleurodon*, from England. Both of these reached lengths of 13–14 metres and had large heads nearly 3 metres long, armed with long curved teeth, some nearly 25 cm in length. Such beasts were clearly top-line predators that preyed, probably, on other marine reptiles, like their slender-necked cousins, as well as on fishes and large marine invertebrates.

However, recently an even bigger marine killer has been discovered. At the meeting of the Palaeontological Association at the University of Birmingham in England in late 1996, Colin McHenry of the University of Queensland, and his English colleagues Arthur Cruickshank, David Martill and Leslie Noe, presented news of a giant pliosaurid from the 160 million-year-old Late Jurassic Oxford Clay. This gargantuan beast of the sea had neck vertebrae 40% larger than those of *Kronosaurus*. Such a megapliosaur may have reached lengths of 18–20 metres and weighed up to 50 tonnes. According to Bob Bakker of the University of Colorado at Boulder, pliosaurids had jaws four times more powerful than those of tyrannosaurids, due to the massive jaw musculature that inserted on them[3]. There is little doubt that they were capable of killing anything in the Mesozoic seas, and could easily snap off the head of their slender-necked plesiosaurian relatives.

The story of *Kronosaurus* goes back to the 1930s, but newly discovered material is revealing more about this huge pliosaur. For a long time *Kronosaurus* held the mantle of being the largest known relatively complete skeleton of a giant pliosaur. It was dug up in northern Queensland in the 1932–33 expedition led by William Schevill of Harvard University. The skeleton was blasted out of the rock with dynamite by a field technician known affectionately as 'the maniac' for his love of blowing the hell out of things. Then the large rocks containing the bones were wrapped in bloodied sheep fleeces to protect them on the long trip back to the USA. On their arrival the horrified director of the Museum of Comparative Zoology at Harvard University had all the specimens promptly disinfected and the fleeces destroyed, for fear of bringing the disease anthrax into the country! Then followed the slow preparation of the skeleton, with minor interruptions, such as World War 2, holding up the work. In short it took nearly 25 years from the time of its discovery until the skeleton, complete with missing bits

No, it's not the Loch Ness Monster! Just a reconstruction of a short-necked plesiosaurian, called a pliosaur, one of the many forms of marine reptiles that lived during the Mesozoic Era. New fossil remains indicate that a huge 'megapliosaur' living in England about 160 million years ago may have reached 20 metres long.

filled in by plaster, went on public display at Harvard's Museum of Comparative Zoology in 1959[4].

Despite being one of the most complete large pliosaurids known, the skeleton has never been formally described. The Queensland Museum palaeontologist, Heber Longman, originally named the beast after a small fragment of the snout that he had studied back in 1924, and today this remains as the type specimen. Longman had no formal training in science, and was self-taught. However, he did an excellent job of describing many of Australia's dinosaurs, ichthyosaurs, plesiosaurs, turtles and even mammals. His wry sense of humour comes out in the naming of *Kronosaurus* after the mythological Greek God *Kronos*, who ate his own children.

However, there is a renaissance in the study of *Kronosaurus*. Colin McHenry at the University of Queensland is presently revising the genus, on the basis of much newly collected material. Other specimens have been collected by expeditions from the Queensland Museum and the University of Queensland over the last few decades. Amongst this material is a new skull, which indicates that the one restored for the Harvard skeleton may not be entirely accurate. Some palaeontologists have suggested that there may be two species of *Kronosaurus*, perhaps even two separate genera. The new material studied by McHenry will hopefully sort out the identity crisis of *Kronosaurus* and enable its relationships to other pliosaurids to be resolved.

There have been many new discoveries and a plethora of new work carried out on Australia's and New Zealand's fossil marine reptiles over the last few years. Perhaps the most notable is that of Eric. This magnificent opalised skeleton of a small pliosaurid was found at Coober Pedy in South Australia. Originally dug up by an opal miner, whose digging machine churned right through the skeleton, the remains were then purchased by entrepeneur Sid Londish, who paid the Australian Museum to have it restored and fully prepared. This labour of love took palaeontologist Paul Willis some 450 hours of work. But the results are stunning and worth the long hours of toil.

What emerged was an almost complete skeleton, including a well-preserved skull. The bones had flashes of colour because the entire skeleton was preserved as precious opal. And herein lay the dilemma. A new skeleton, something that the palaeontologists suspected was new to science, yet the sheer value of the opal content of the specimen made it extremely valuable. What was to become Eric's fate? The story took a turn for the worse when Eric's owner became bankrupt and the banks seized his assets, including Eric. They saw Eric as a way of realising some of his assets, so they immediately put the magnificent specimen up for public auction. At the time there were no effective laws in place to prevent Eric from being lost overseas. Since then the Protection of Movable Cultural Heritage Act has come into existence, one of its aims being the protection of the nation's fossil heritage. Likewise, there were no laws in place that would prevent someone within Australia from buying it and then cutting it up into little pieces for sale. Fortunately, the scientific value of the specimen won through, although this also had a monetary basis as Eric as pure opal was only worth about Australian $100 000, yet as an opalised pliosaur he was deemed to be worth at least $300 000.

The problem, however, was still how to come up with that sort of money in order to save it for the people of Australia. Then in stepped Alex Ritchie, Curator of Palaeontology at the Australian Museum. Alex had one of his frequent brilliant ideas and proposed that a scheme should be launched on national television. So, the popular science programme 'Quantum', that just happened to be hosted by a long-time friend of Ritchie's, Karina Kelley (they had both emigrated to Australia many years before on the same ship), put out a plea for donations to save Eric. The response was phenomenal. Within three weeks enough donations had poured in from around the country from individuals and companies, enabling Eric to be purchased by the Australian Museum, at a cost of around $340 000.

So, after all that what of the scientific status of Eric? As he had become such a popular character, he became the focus of much media attention and was immediately put on travelling exhibition around Australia so that everyone who contributed towards his purchase could get to see him. His long trip around the country took about two to three years. Eric finally came to rest at the Australian Museum where work continues on uncovering his identity.

A series of fortunate events and expeditions by Arthur Cruikshank of the Leicester Museum in England has done much to work out not only what Eric is, but details of his close relatives and their wide distribution in Cretaceous times. In the early 1990s another find was made that was to help elucidate Eric's relationships. A series of fossil-collecting expeditions around the Kalbarri region in Western Australia uncovered some partial skeletons of a small pliosaur, of similar age to Eric but slightly bigger and more robust. In 1996 Cruickshank visited Western Australia and we began to work on the description of the new material. These were the first articulated plesiosaurs that had ever been found in the western half of the continent. The year before I had met Arthur Cruickshank for the first time at a palaeontological meeting in South Africa, where he had stayed on to study one of South Africa's best preserved plesiosaurs, a small animal discovered near Port Elizabeth, and known at that time as *Plesiosaurus capensis*.

As a result of his visit to Australia, Cruikshank realised that the Australian and South African specimens, along with another from England, were all closely related[5]. He had been studying the remains of a small pliosaur known as *Leptocleidus superstes* for some time in England. On going to South Africa Cruikshank recognised that *'Plesiosaurus' capensis* was actually another species of *Leptocleidus*. On coming to Australia he discovered yet more *Leptocleidus* – Eric and the Western Australian plesiosaur were also both examples of new species of *Leptocleidus*[4]. Thus this relatively small seal-like plesiosaur, about 2 metres long, descended from large 5 metre-plus early Jurassic plesiosaurs, was suddenly world-wide in its distribution in the early Cretaceous, about 100 million years ago, from Europe to South Africa and Australia.

New Zealand also has a rich record of marine reptile discoveries, ranging back to the early nineteenth century when a dynamic period of active collecting was undertaken by a number of museums to procure specimens. Although many of these are still in New Zealand or England today, some, unfortunately, were lost. For instance, when the *Matoaka*, set sail bound for England in May 1869, carrying fossil marine reptiles for Richard Owen at the British Museum, it was never seen again. The reptiles had returned to a watery grave[4].

In recent years Ewan Fordyce of Otago University has been the driving force behind some of the major new discoveries in New Zealand.

One spectacular discovery is the famous Shag Point plesiosaur, an almost complete skeleton of a 7 metre long animal excavated by Fordyce and his colleagues in the early 1980s from a nodule exposed on the beach just north of Otago. 'Shaggy', as he is affectionately known, represents another new genus that is currently being described by Fordyce and Cruickshank. At this stage their research has revealed that this plesiosaur appears to be closest to others in a family, the cryptocleidids, that are mostly known in the Late Jurassic of Europe and North America. The Shag Point animal is a highly derived cryptocleidid in having a much longer neck (with more vertebrae) than a normal cryptocleidid, yet retaining a regular neck vertebral morphology for the group. Yet, it is Late Cretaceous in age, suggesting a line of independent evolution for the family in the Southern Hemisphere, long after they had been displaced by newer groups in the Northern Hemisphere. Its affinities lie more with other plesiosaurs known from Chile, Argentina and Antarctica.

A new interpretation of the patterns of extinction and replacement of pliosaurids, plesiosaurids and ichthyosaurs across the Jurassic/Cretaceous boundary has been proposed by Bob Bakker[3]. According to his model, the long-necked plesiosaurians of the Cretaceous seas are not, as you would expect, descendants of the long-necked plesiosaurians of the Jurassic. Rather, he argues, they evolved independently from the pliosaurids which survived the end of the Jurassic. Based on the detailed comparative anatomy of the skull, Bakker has suggested that the true plesiosaurs died out at the end of the Jurassic, and a new radiation from pliosaurids filled in the gaps left by the extinction.

Bakker has argued not only that giant pliosaurids continued through the boundary unaffected, but from them evolved the long-necked elasmosaurs, some of which reached huge lengths of up to 15 metres. Smaller, long-faced polycotylids, like *Dolichorhynchops*, he suggests, would have evolved to fill the niche left vacant by mid-sized ichthyosaurs. The smaller pliosaurids, like *Leptocleidus*, appeared at this time and occupied a niche similar to that of modern seals, living near-shore and being short-ranging fish eaters. The Shag Point plesiosaur, a possible late-surviving cryptocleidid, can now be seen in a more interesting light. If 'Shaggy' is indeed a cryptocleidid and Bakker is right, it implies that some of the more primitive, late Jurassic plesiosaurians failed to survive the Jurassic/Cretaceous boundary, but only in the Southern Hemisphere. Exactly why this would have happened is unclear, because large, killer pliosaurids were still lurking in the southern oceans at this time just as they were in the Northern Hemisphere.

In addition to its fine record of plesiosaurs, New Zealand also boasts a superb record of the giant sea-going lizards, the mosasaurs[4]. They are represented by well-preserved skulls and partial skeletons. Mosasaurs were one of the earliest groups of fossil marine reptiles ever to be discovered, the first having been found about 40 years before the first dinosaur. This was the famous Maastricht mosasaur, uncovered in Holland near the town of Maastricht back in 1780. Following its discovery the specimen became the subject of a legal battle over its ownership between Monsieur Hoffman, who oversaw the excavation, and the Canon Goddin, on whose land the specimen was found. (Things haven't changed much – a similar battle has ensued in recent times over the ownership of a *Tyrannosaurus rex* skeleton known as Sue, between the excavator and the Federal Government on whose land it was found.) When the French army beseiged Maastricht in 1795 the house where the mosasaur specimen was kept was not fired upon. Suspecting that the French knew of the specimen's whereabouts, the owner had it hidden in a secret vault in the township. The French authorities certainly did know about the famous fossil and wanted it. So they offered 600 bottles of wine for its recovery! Not surprisingly, it was found soon after, and today the specimen resides in Paris, where it is on display in the Natural History Museum in Paris. The specimen was later given the name *Mosasaurus*, meaning 'lizard from near the Meuse River'.

Mosasaurs were very lizard-like in appearance with very long, powerful tails. They swam like crocodiles using their tails to propel them through the water, aided by their powerful flippers. They grew to enormous sizes, the biggest being around 15 metres long. The group is exclusively confined to Late Cretaceous rocks. New discoveries made in New Zealand show that, with a few exceptions, Southern Hemisphere mosasaurs mostly belong to the subfamilies that lived in the Northern Hemisphere[4]. Thus genera like *Mosasaurus, Tylosaurus* and *Prognathodon* all occur in New Zealand as well as in North America and Europe.

Mosasaurs have a long skull armed with many large, curved teeth. The architecture of the skull is built around a central hinge and the lower jaw was also capable of hinged movement. On first examining the deadly looking teeth of a mosasaur one would assume that they were voracious predators, capable of tearing apart other large marine reptiles or big fishes. However, despite some fish remains being found inside some mosasaur skeletons, several other interesting feeding adaptations for mosasaurs are evident. For example, in 1960 it was shown that a certain ammonite, *Placenticeras*, was often found with large rows of

Would you pay 600 bottles of wine for this specimen? Someone did. It is the famous Maastricht mosasaur, discovered in Holland in 1780, and was the centre of the first legal wrangle over the ownership of a fossil. The invading French armies then tried to find the controversial specimen and so put up the reward to uncover it. They did, and it still resides in the Natural History Museum in Paris to this day.

tooth-bite marks across the shells[6]. These match precisely the outer and palatal tooth patterns, and are of the exact same size, to correspond with the jaws of a mosasaur. In some cases the reptile had attacked the poor cephalopod 16 times before it met with success in cracking open the shell.

I was present at the DinoFest 2 meeting in Tempe, Arizona, in 1996 when a hot debate about mosasaur feeding habits developed immediately after a talk on their feeding strategies by Erle Kauffman of the University of Colorado. Kauffman had shown that mosasaurs had several feeding strategies when preying on ammonites. These included coming up from behind them, and repositioning them in the mouth so as to get the head and body chamber in the jaws, and even ripping the animal out of its shell by vigorous back and forth shakes of the head. Immediately after Kauffman's talk some scientists argued that ammonite shells would not crack with nice, neat tooth holes when bitten, but should shatter; whereas others argued that the mosasaur was biting so swiftly that the punctures were self-evident from their bite marks. Suddenly Bob Bakker got up and said that they hadn't considered the mosasaurs' special adaptation for biting – the hinged jaws.

Demonstrating with a loud slap of his arms, Bakker then showed that mosasaurs didn't crunch swiftly on the ammonite prey, but instead used its hinged jaws to wrap around the shell and apply even pressure to break open the living chamber where the juicy flesh was. Most agreed with this interpretation.

Other mosasaurs like *Globidens* had hemispherical rounded nut-cracker-like teeth. Although early speculation puts these down to having been used to crack open oysters or hard clam shells, to me they seem to have been equally well-suited for the crushing of ammonite shells. The rounded teeth give greater surface pressure, and thus could be applied to harder-shelled ammonite species. This would have enabled mosasaurs to crush effectively large ammonite shells and pull out the tasty flesh from inside. Yet there is direct evidence that suggests some mosasaurs ate other marine invertebrates. One specimen of the lower jaws of the mosasaur *Compressidens* had a crushed sea-urchin test stuck between its teeth, and, similarly, a specimen of *Plioplatecarpus* had belemnite endoskeletons (from inside a squid-like animal) found between its lower jaws and near its neck vertebrae[7]. If we look at modern marine animals today, there are some parallels. *Physeter*, the sperm whale, armed with many peg-like pointed teeth, feeds primarily on squid, as does the dolphin *Globicephala*. Perhaps mosasaurs were just too clumsy at swimming to chase down large fishes on a regular basis, or catch their reptilian brethren, so they may have rapidly adapted to a niche of feeding primarily on invertebrates.

Recent research on mosasaur origins and relationships is unravelling a more complex story than had hitherto been supposed. For many years mosasaurs were thought to be very closely allied to the varanid lizards and were classified accordingly in the Superfamily Varanoidea. Varanids include the many forms of goannas in Australia, as well as the world's largest living lizard, the 4 metre long Komodo dragon of Indonesia. Snakes were thought to have evolved independently from lizards some time during the Cretaceous, by a paedomorphic reduction and ultimate loss of limbs. This is likely to have been as part of a developmental trade-off that saw a dramatic increase in the number of vertebrae. In recent years two important scientific papers have been published that propose a radical new model, based on fossil evidence, for the origins of snakes and the evolution of mosasaurs.

The first piece of research by Michael Caldwell of the Field Museum, Chicago, and Michael Lee of Sydney University, presented a new interpretation of *Pachyrhachis problematicus*, a slender mid-

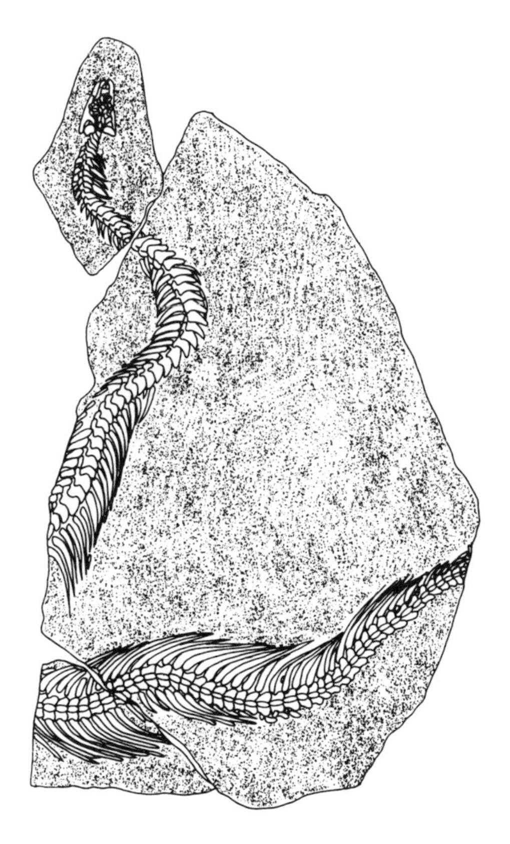

Is this the original serpent from the garden of Eden? Pachyrhachis *is the most primitive fossil snake known. It hails from the Late Cretaceous of Israel. It still retains small hind legs (although these are not preserved on this particular specimen). Snakes are now thought to have evolved in the sea and to be closely related to the mosasaurs.*

Cretaceous reptile from Israel[8]. This had previously been thought to be a varanoid lizard. By carefully re-examining the specimen, Caldwell and Lee identified it not as a lizard, but as a snake, based on 19 advanced features of the skull. The skull is highly kinetic, that is to say that the palatal bones can move to a moderate degree, enabling the mouth to open very wide. *Pachyrhachis* is found in marine deposits indicating that it inhabited a shallow sea environment. Caldwell and Lee also identified very small hindlimb elements on the two known specimens, showing that although the front limbs had been lost by that time, rudimentary hind legs were still present.

The loss of digits on the foot is quite clear. Snakes are normally thought of as having been derived from lizard-like ancestors which then

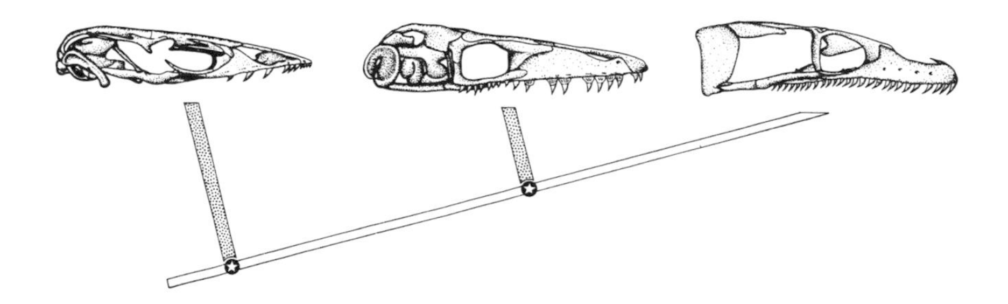

The idiot's guide to snake origins. This incredibly simple diagram shows the latest theory in the evolutionary relationships between varanid lizards (left), mosasaurs (middle) and snakes (right). We blame Mike Lee *for this.*

subsequently lost the limbs. The way in which they lost their limbs has been a major point of contention. The new interpretation suggests quite strongly that snakes evolved from reptiles that were ancestral to mosasaurs, and that they lost their limbs through adaptations to swimming in water. The main form of propulsion for mosasaurs and snakes is the same – using the long, powerful tail, rather than using flippers, as do plesiosaurians.

Mike Lee continued this idea in a second paper that he published, in which he discussed the phylogeny of varanoid lizards and the origins of snakes at some depth[9]. As in his paper with Caldwell, Lee argues that snakes are actually more closely related to the sea-going mosasaurs than they are to varanoid lizards. He unites mosasaurs and snakes in a group called the Pythonomorpha, and supports this conclusion with a rigorous analysis, supported by the presence of nearly 40 shared characteristics that are only found in snakes and mosasaurs. The discovery of the true affinities of *Pachyrhachis* was crucial to this interpretation. So the picture emerges of snakes probably having gained their long, slender bodies and reduced limbs within the aquatic environment, and then having completely lost their limbs, conquering the terrestrial habitat.

What comes out of Lee's detailed analysis is that mosasauroids, the group containing mosasaurs and their immediate ancestral relatives, are the nearest relatives of snakes. True terrestrial snakes appear by the Late Cretaceous, at the same time that the radiation of mosasaurs was peaking. For many years the origin of snakes was clouded in mystery due to lack of adequate fossil evidence. But now Lee and Caldwell have blown away the cobwebs, by taking a fresh, rigorous, and very detailed, look at the known material. They have come up with a quite novel interpretation. Whether their ideas will stand the test of time remains to be seen.

So when you are next out snorkelling and are startled by a sea snake, it may not only be some highly derived snake that you are frantically paddling away from, but all that remains of a great radiation of aquatic reptiles that once dominated the seas. Their place has now been taken by the cetaceans: the ichthyosaur-like dolphins, and the pliosaur-like toothed whales. But don't stare at the snake in silent awe for too long, as a sea snake's bite is 10 times stronger than that of a cobra; so get out fast, and then reflect on the wonders of evolution.

11

Reign of Pterosaurs

The first flying vertebrates

'Well, suddenly out of the darkness, out of the night, there swooped something with a swish like an aeroplane. The whole group of us were covered for an instant by a canopy of leathery wings, and I had a momentary vision of a long snake-like neck, a fierce red greedy eye, and a great snapping beak, filled to my amazement with little gleaming teeth. The next instant it was gone – and so was our dinner. A huge black shadow, twenty feet across skimmed up into the air; for an instant the monster's wings blotted out the stars, and then it vanished over the brow of cliffs above us.' What Ed Malone, an adventure-seeking character from Conan Doyle's 1912 novel *The Lost World*, was accurately describing was one of the extinct flying reptiles that lived in the age of the dinosaurs.

Coincidentally, the setting for the 'lost world' is ambiguously described as in a high plateau in the Amazon jungle in Brazil. Conan Doyle's words were to be prophetic because 60 years after the publication of *The Lost World*, pterosaur bones were found on the Araripe Plateau in Brazil, and to date this region is one of the world's most important Cretaceous fossil sites for pterosaurs, due to their beautiful preservation.

Pterosaurs have both fascinated and baffled people since their first discovery more than 200 years ago[1]. The earliest description of one was in 1784 by the German naturalist Cosimo Collini. He examined a beautifully preserved specimen of *Pterodactylus antiquus* from the famous Solnhofen quarry at Eichstätt and recognised that the creature had wings. Yet he did not place it with the birds, instead concluding that it was impossible to place into any of the known groups of animals. His accurate reconstruction of the fossil later played an important role in elucidating the true nature of the pterosaur group. The first person to recognise that the Eichstätt fossil was a reptile was the famous French anatomist Georges Cuvier. Although he never saw the actual specimen, Cuvier was able to deduce from Collini's clear drawing that it was a flying creature, and, in a paper published in 1809, he erected the name 'Pterodactyle' for it, meaning 'flight finger'. The term 'pterodactyl' has now become entrenched in the popular literature for these ancient flying reptiles, but the common word 'pterosaur' (meaning 'winged lizard') is the correct vernacular name for their group, the Order Pterosauria.

The original Eichstätt pterosaur was later restudied by Samuel von Soemmerring of the Bavarian Academy of Sciences, who concluded that it was a bat-like mammal. When a second pterosaur turned up in 1817 from near Eichstätt, this only made Soemmerring more strongly convinced that pterosaurs were mammals. Few of the eminent naturalists of the day supported Cuvier's notion that they were reptiles, a conclusion he had based on his sound knowledge of reptile skeletal anatomy. The view that pterosaurs were really reptiles was not widely held until the twentieth century, despite a handful of good anatomists that insisted they were reptiles.

Harry Glover Seeley of King's College, London, insisted in the mid-nineteenth century that pterosaurs must have been warm-blooded, and were probably intermediate between the reptiles and birds, being more like dinosaurs. The famous anatomist Sir Richard Owen violently disagreed with Seeley, seeing pterosaurs as cold-blooded reptiles incapable of high metabolic flight, which must, he at first predicted, limit their maximum size. Up to that point no large pterosaurs had been discovered. Then, in 1846 a large pterosaur, having an estimated wingspan of about 5 metres, was described from the Chalk in Kent. In answer to this Owen simply said that God's power as a Creator had been greater in the past and thus went beyond his calculations.

In 1870–71 the famous American dinosaur hunter Othniel Charles Marsh discovered the first pterosaurs from the USA. These were gigantic animals, having wingspans of the order of 6 to 7 metres. Marsh later named these flying behemoths *Pteranodon* (literally meaning 'winged without teeth'), as the huge beak had no teeth. This separated it from the toothed genera then known from Europe. For many years

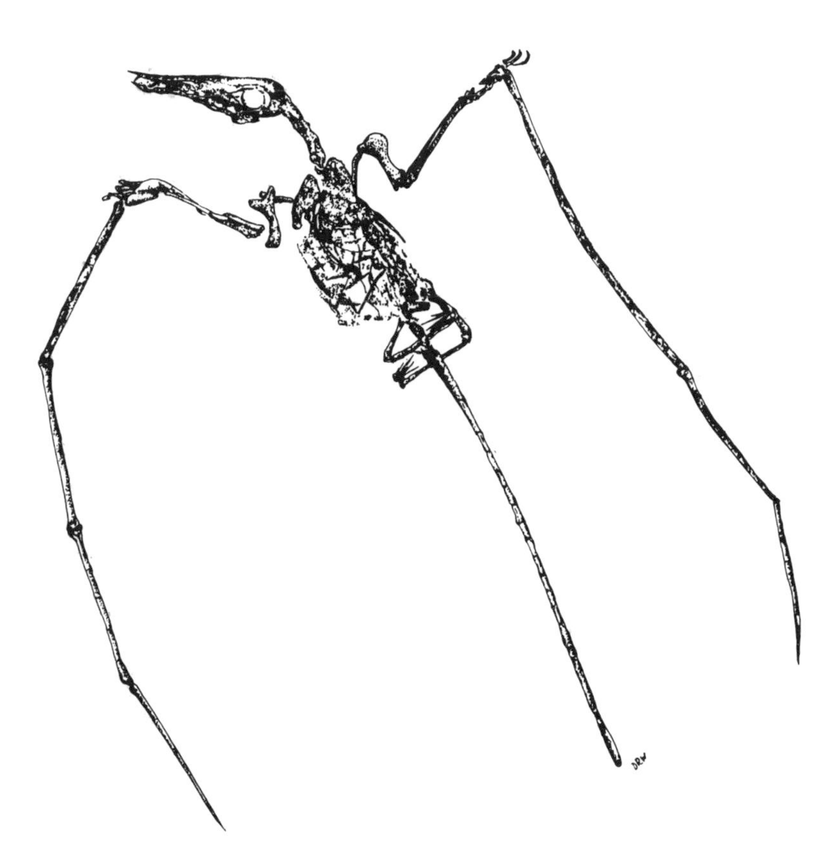

A bat out of hell? No, just the fossilised remains of a pterosaur from the Late Jurassic Solnhofen Limestone of Germany. The first such fossils were found and studied from these rocks, and caused much consternation amongst early zoologists, who couldn't agree on whether to place them as birds, bats or reptiles. Pterosaurs dominated the skies of the Mesozoic Era, but died out alongside the dinosaurs at the end of that time.

Pteranodon held the crown as being the largest known pterosaur, until 1975 when remains of an even bigger beast were found in Texas by Doug Lawson and his mentor Wann Langston Jr, both of the University of Texas. Over the last 25 years more of this beast, named *Quetzalcoatlus northropi*, was uncovered by Langston's expeditions, enabling a more or less complete reconstruction to be made of the largest flying animal the world had ever seen.

Quetzalcoatlus was as large as a small aeroplane, having an 11–12 metre wingspan. It also had an amazingly large body, maybe 8–10 metres from toes to snout. Yet recent estimates suggest that the whole animal may have weighed as little as 86 kg[1]. The lifestyle of this enigmatic giant has been postulated as having been anything from like a giant, scavenging vulture, preying on the carcasses of dinosaurs, to an invertebrate feeder that poked its beak deeply into muddy burrows in search of little crustaceans! Only recently, in mid-1996, has the actual head of *Quetzalcoatlus* been described by Alexander Kellner of the American Museum of Natural History and Wann Langston Jr[2]. They showed that the head had a long, pointed narrow beak and bore a rectangular crest above the huge nasal–preorbital opening, so the eye was positioned quite low on the head. We do not know the extent of the median crest above the eyes, but presume it functioned to counterbalance the weight of its long snout, which was about 1 metre long, or to give aerodynamic stability whilst in flight.

The fact that the snout was so long led Tom Lehman, of the Texas Technical Institute in Lubbock, and Wann Langston to propose a radical, new interpretation of the lifestyle of *Quetzalcoatlus* at the 1996 meeting of the Society of Vertebrate Paleontology in New York[3]. They claimed that it fed on invertebrates by probing its long beak into sediments. This idea was based solely on the circumstantial evidence that the bones had been found in sediments that contained burrows that might have been made by crustaceans. This idea was picked up by journalists and soon became entrenched in the popular science media, despite criticisms by some other palaeontologists that the idea was not based on any sound evidence. No stomach contents had been found, and this is the only direct means of interpreting the diet of extinct animals. Indeed, several other pterosaur fossils have been found with remains of fishes in their gut region.

In the case of *Quetzalcoatlus,* and indeed most fossil vertebrates, the place where they finally became buried may in no way reflect where they lived their lives. Many terrestrial dinosaur remains are found in marine deposits as the carcasses float out to sea and eventually sink. Similarly, bloated cow carcasses today can be found out at sea after severe flooding. This does not indicate that cows normally dive for fish.

David Martill of the University of Portsmouth, England, has criticised the way that the science media can jump to sudden conclusions without any evaluation of a new idea, and make it an accepted printed 'fact'[4]. Martill has his own idea about what *Quetzalcoatlus* ate. He thinks that it was a fish eater. *Quetzalcoatlus* had an exceptionally long neck and skull, the two combined measuring nearly 5 metres from snout to shoulder. The elongate, rigid nature of the neck vertebrae indicates

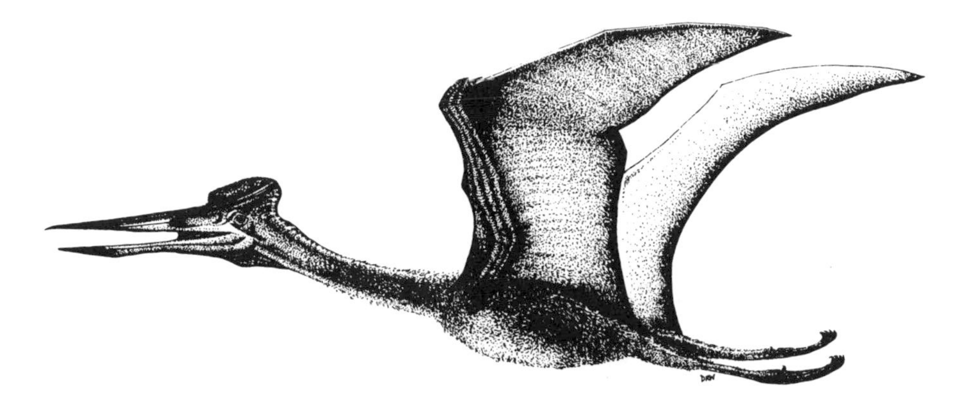

You wouldn't want to have sat under this baby's tree. The largest creature ever to fly was this pterosaur, Quetzalcoatlus, *whose 12 metre wingspan approximates that of a small airplane, and whose droppings would make more than a white speck on your hat. Despite its huge size, it was most likely a fish eater whose size was advantageous in allowing it to scoop up fish whilst keeping it a safe distance above the water level.*

that it had a stiff neck that could be lowered like a rod when it was gliding about 3 metres above the water level. Like this it could easily have lowered its huge head to scoop up fish. As it had a very tiny throat, small fish would seem to have been its most likely food source. Martill explains that the huge size of *Quetzalcoatlus*, in particular its long neck, were probably adaptations that enabled it to stay a safe distance above the water level for gliding, without any risk of wetting the wings and becoming submerged. It probably would have had great difficulty taking off if it got waterlogged.

The seas of the time were a dangerous place, being inhabited by many large reptilian predators (see Chapter 10), such as mosasaurs and plesiosaurs, so keeping a safe distance above the water level was imperative for a pterosaur. The stiff neck of *Quetzalcoatlus* was also a good reason why it couldn't have fed like a vulture. Vultures and other carrion-feeding birds require some dexterity in the neck to both twist and pull out pieces of flesh and viscera. The slender shape, elongate feeble legs and long pointed beak that lacked a hooked end, all support the notion that this gargantuan flyer fed upon small fishes.

Underlying the widely differing interpretations of lifestyles, such as in the case of *Quetzalcoatlus*, is a factor that has bugged palaeontologists for a long time: could pterosaurs really fly, or were they just gliders that were at the mercy of the winds? The other age-old question of how competent they were on their feet goes hand in hand with their supposed flying abilities.

For many years the traditional view of the flying ability of pterosaurs was that because they were thought to be cold-blooded reptiles they must have lacked the metabolism needed for true flight, as today only warm-blooded creatures like birds or bats can generate enough energy to sustain true flight. This is, of course, the very same reason that many early workers on pterosaurs saw them as having to be warm-blooded, and thus more closely related to mammals or birds. In recent years the debate has taken a sharp turn around due to discoveries over the last decade or so that dinosaurs may have been partly or fully warm-blooded. As we now know, pterosaurs are most closely related to the common ancestors of dinosaurs. It follows that they may well have had a very similar physiology.

The first discovery of pterosaurs having dense fibres on their wing membranes was made by German palaeontologist Karl Alfred von Zittel in 1882[1]. Ferdinand Broili in 1927 interpreted needle-like dots on the wing membranes as hair follicles, which seemed to suggest a warm-blooded affinity to the mammals. But the most exciting discovery was published in 1971 by Moscow palaeontologist A. Sharov. He announced the discovery of a new pterosaur covered in fur, or at least small, fur-like fibres, from the Karatau Mountains of Kazakhstan[5]. Named *Sordes pilosus*, meaning 'hairy evil spirit', this creature was to fan the flames of pterosaur controversy for another decade. The impressions of fibres in the wings of pterosaurs had previously been known from well-preserved specimens from Germany, but *Sordes* showed that actual fur-like fibres covered the whole body, and enveloped the legs, leaving the naked tail free to act as a rudder. This immediately pointed to pterosaurs as being most likely a warm-blooded group with a high enough metabolism to undertake self-propelled flight, as opposed to passive gliding.

In a study of *Sordes pilosus*, David Unwin and Natasha Bakhurina of the University of Bristol showed that the exceptionally well-preserved wing membrane was divided into two regions: a stiffened outer half and a softer, pliable inner half[6]. The hindlimbs were attached to the main wing membrane, and thus took an important role in the flight apparatus. Their detailed study contradicted suggestions of earlier workers who had argued that the extensive furry covering was restricted to the wings, and not seen as a displaced sheet of skin. Thus the so-called fur of *Sordes* is now seen as a part of the specialised close-packed fibres that supported the wing membrane, and their different densities in the wing reflect the animal's ability to undertake slow, manoeuvrable flight. This would have been imperative for fish- or insect-eating flying creatures.

The new interpretation of Unwin and Bakhurina also has important implications for the terrestrial locomotory abilities of pterosaurs. Whereas some workers, like Kevin Padian of the University of California, Berkeley, have argued that pterosaurs had no wing membrane covering the legs, and thus were good walkers and runners on land[7], these new finds suggest otherwise, at least in *Sordes*. The presence of wing membranes covering the legs would suggest that, like bats, pterosaurs at best could only crawl along the ground and were more at home roosting up trees or on cliffs. Furthermore, if all pterosaurs had the same method of flight, with legs enveloped by wing membranes, the origins of pterosaur flight would be most likely to have been derived from climbing and gliding (i.e. gravity assisted) rather than from fast-running beasts taking progressively longer and higher leaps into the air, which requires well-developed free legs. In other words, they flew from the tree, or cliff, down, rather than from the ground up, as has been suggested for the evolution of flight in birds (see Chapter 13).

The debate on pterosaur flight and terrestrial locomotion does not end with Unwin and Bakhurina's study however, as some vital pieces of evidence were not considered, notably the trackways of pterosaurs. Well-preserved fossil trackways are known that have been identifed as pterosaurs based on their distinctive foot structure, coupled with the rare association of hand and wing impressions. Such trackways have led Chris Bennett of the University of Kansas to propose that some of the rhamphorhynchid pterosaurs had erect quadrupedal locomotion, and that some of the larger pterodactyloids had upright walking bipedal locomotion, as argued by Kevin Padian since 1983[7]. In a recent study Chris Bennett reanalysed the trackways known as *Pteraichnus*, and concluded that they were made by a pterosaur moving at about 1 metre per second, using postures that varied from bipedal to sprawling, semi-erect gaits[8]. Furthermore, Bennett's enthusiasm to demonstrate his point is shown by his construction of a model of a *Pterodactylus* skeleton with movable limbs, which demonstrated how his proposed method of pterosaur terrestrial locomotion was possible. Bennett's refined model now sees pterosaurs as being capable of swift running to gain momentum before leaping into flight, or else using their powerful hindlimbs to leap high into the air and initiate flight.

As we shall see, not all pterosaurs were built along the same lines, in terms of differing wing to body length ratios and feeding mechanisms, so differing flight capabilities almost certainly existed across the diversity of pterosaur families. Bennett concluded that pterosaurs were capable of efficient bipedal locomotion, but for feeding and slow walking a gait incorporating the hands was preferable. But maybe not all pterosaurs behaved in the same way. To appreciate the great diversity of lifestyles that pterosaurs adopted in the Late Mesozoic Era we need to take a look at pterosaur evolution to gauge the nature of this great aerial reptilian radiation and the breadth of diversity that this group achieved before succumbing to the same grisly fate as the dinosaurs.

Pterosaur origins are still a mystery. Some workers, including Kevin Padian, have argued that on the basis of foot structure pterosaurs evolved from thecodonts, the reptilian group that is considered most likely to have also given rise to dinosaurs[9]. Other palaeontologists, like pterosaur expert Peter Wellnhoffer of the Natural History Museum in Munich, consider eosuchian reptiles, lizard-like beasts, to be better contenders as ancestors of pterosaurs. As the first pterosaurs appeared as complete pterosaurs, not showing any intermediate features, it appears that the group evolved rapidly, and was established by the beginning of the Late Triassic, about 230 million years ago.

The oldest known pterosaur is a small form, *Eudimorphodon ranzii*, from northern Italy. It had a wingspan of about 1 metre, and a body length from snout to feet of about 40 cm. Despite its primitive reptilian features, the jaws show specialised teeth in having enlarged fangs present in the middle of the jaws and at the front of the mouth. Some of its teeth are multipronged, showing specialised 'molar'-like shapes. It is believed that the teeth wore down from its habit of biting the hard scales of the fish that it was preying upon, as fish scales have been found in the fossil in the region where its stomach would have been. It has been suggested that the juveniles may have had a different dentition that was more suited to catching insects.

Several of the other early pterosaurs had a similar body form to *Eudimorphodon*, such as the pigeon-sized *Preondactylus* and the owl-sized *Peteinosaurus*, both of which also come from the same beds as *Eudimorphodon*, but are slightly younger in the section. *Preondactylus* appears more primitive than *Eudimorphodon* in having a less specialised dentition, but is otherwise very similar in general shape. *Peteinosaurus* has a deeper snout than the others, with jaws having many small teeth, and only a few enlarged teeth, at the front of the mouth. All three of these early pterosaurs appear to have been adapted to catching fish.

Many of the known Jurassic pterosaur fossils have come from the famous Solnhofen Limestone in Germany. These were all Late Jurassic in age, about 150 million years old. Early–Middle Jurassic pterosaurs are rare, but some interesting forms have been found from the Lias beds of England, that range in age from about 195 to 215 million years ago. One of these, *Dimorphodon*, has a remarkably deep head with enlarged teeth at the front and smaller back teeth, still showing the sorts of dental adaptations suitable for a fish diet, bearing in mind the hard-scaled types of fishes that were around then.

The first pterosaurs that exhibited quite specialised feeding mechanisms come from the Early Jurassic strata of Germany. *Dorygnathus*, with a 1 metre wingspan, had a skull sporting very enlarged projecting front teeth, and tiny rear teeth, an effective mechanism for gripping and holding slippery food, like fishes. *Ctenochasma*, from the Late Jurassic of Germany, was a short-tailed form with many long, tightly packed, yet widely splayed teeth in the jaws, probably for comb-filtering the water for small organic invertebrates. *Anurognathus*, also from Solnhofen, had an extremely deep head with few, short, peg-like teeth widely spaced on the jaws. It has been suggested by Wellnhoffer that it fed on insects, and thus must have been a very manoeuvrable flyer[1].

Dsungaripterus, from the Early Cretaceous of China, is one of the most bizarre looking beasts to ever take to the skies. Its skull bore a long median crest, as did many other pterosaurs, but the upturned beak was lined with rounded knobbly teeth. It seems as if it was adapted for using its beak-like tweezers to pry out small, hard-shelled invertebrates, then crack them between the rounded teeth. Another form, *Tropeognathus*, from the Early Cretaceous of Brazil, had an enlarged crest of bone over the snout and under the lower jaws, possibly giving it stability as a keel when ploughing the snout through the water while fishing. In contrast, *Pterodaustro*, from the Early Cretaceous of Argentina, had a lower jaw with many hundreds of fine needle-like teeth forming a brush, similar to baleen in a whale. It had only tiny remnant teeth on the upper jaw, so it appeared especially adapted for filter feeding from a standing position, and has been likened to a flamingo of the pterosaur world. Even more bizarre is *Tapejara*, from the famous Early Cretaceous Santana Formation of Brazil, which was a toothless form having a very short, but deep, bill with a high bony crest over the tip of the snout and an enlarged lower jaw crest as well. It is not known what kind of prey it

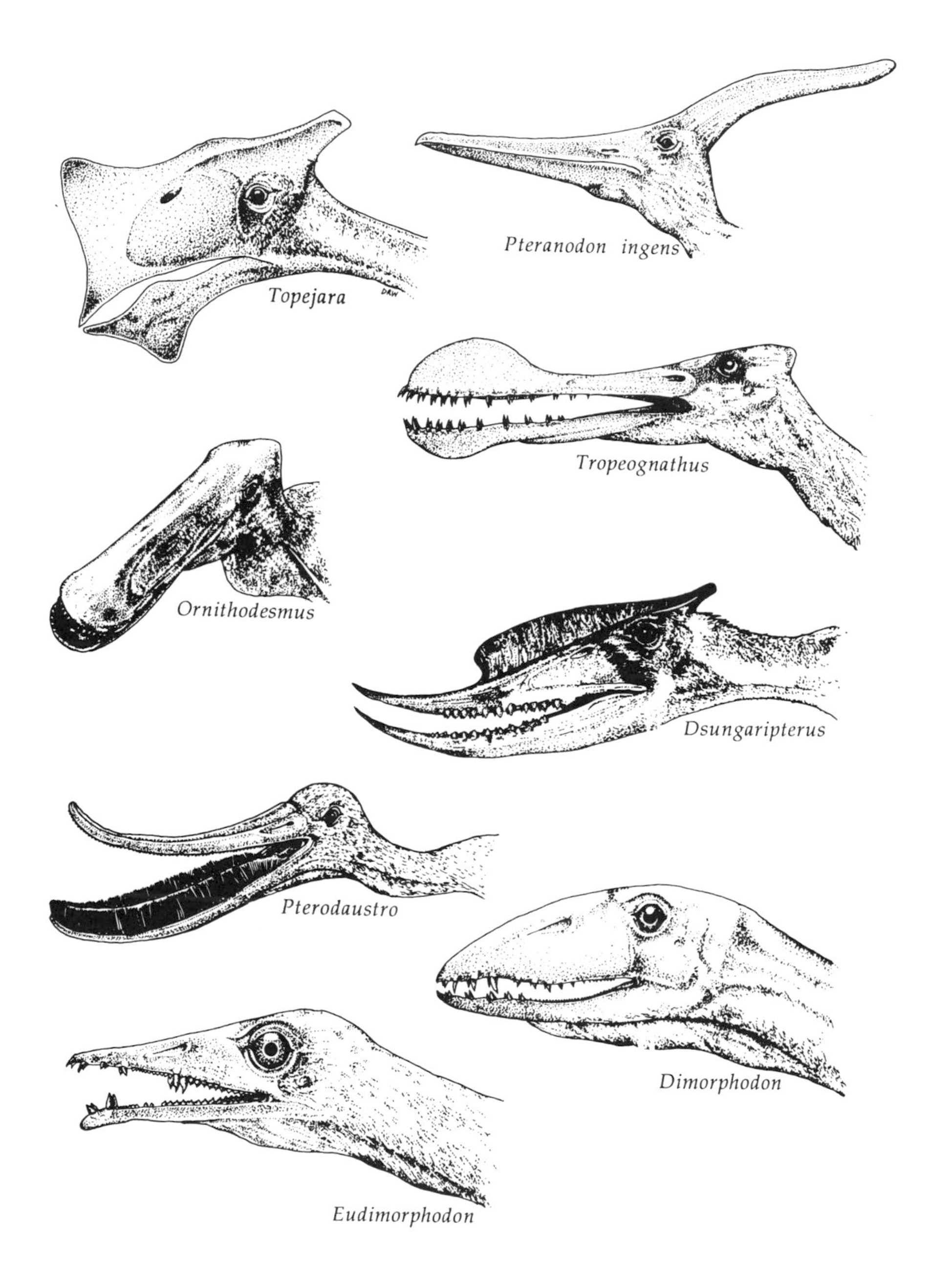

A dentist's nightmare! Pterosaurs underwent a great radiation in the Jurassic and Cretaceous Periods, as shown here by their diverse array of feeding mechanisms. Although principally adapted for eating a variety of fishes, some may have fed on insects and others on hard-shelled invertebrates.

may have fed upon, and as the postcranial skeleton is unknown, no clues are provided as to what the animal as a whole looked like.

Ornithodesmus, from the late Early Cretaceous of the Isle of Wight in England, is one of the few pterosaurs that developed a widely splayed flat snout with a rounded shape, lending it the popular name of 'the duck-billed pterosaur'. Its tightly meshed network of teeth indicate that like other pterosaurs it most probably fed on fish. The most famous pterosaur, *Pteranodon*, had a well-developed bony crest extending from the back of the head, which was developed as a wide flange in *P. sternbergi*, and a more elongate projection in the slightly smaller *P. ingens*. *Pteranodon sternbergi*, had a 9 metre wingspan, whereas it was close to 7 metres for *P. ingens*. Such rudders may have stabilised the head during flight. Interestingly, *Pteranodon* had very thin-walled bones and hollow cavities in the skull to lighten the skeleton. It has been calculated by Cherrie Bramwell and George Whitfield of Reading University, England, that the body weight of *P. ingens* was only about 16.6 kg even though it had a 7 metre wingspan[10]. Such a lightweight body would have enabled the pterosaur to glide for long periods, expending little energy. Such an interpretation fits in well with the discovery of the remains of *Pteranodon ingens* found in marine deposits up to 160 km from the nearest shoreline.

The largest of all pterosaurs, *Quetzalcoatlus*, and its near relatives have recently been discovered from around the globe in deposits of latest Cretaceous age, 65 to 70 million years old. One of these turned up in Western Australia. However, the discovery did not necessitate arduous hours spent trekking through the remote, arid region of this part of the world. When I first took up the position of Curator of Vertebrate Palaeontology at the Western Australian Museum, the first thing I did was to rummage through the fossil collections to look for interesting bones that had not been described. One such specimen came from the Late Cretaceous near Exmouth, in Western Australia. It was in three separate pieces. On gluing it back together I found that it was an elongate bone with very thin walls that were supported by thin cross-struts. I had a cast made and sent replicas to various pterosaur experts. Ralph Molnar from the Queensland Museum was the first to suggest it might be a pterosaur and, on his recommendation, a cast was promptly sent to Peter Wellnhoffer in Munich. He immediately confirmed we had a pterosaur, and a large one at that. Wellnhoffer declined to describe the specimen, which was now identified as the distal end (or hand end) of an arm bone (ulna). I then sent a cast to Chris Bennett in Kansas, and together we described the specimen in 1991[11]. It turned out to be an azdarchid pterosaur, the same family in which the mighty *Quetzalcoatlus* belonged. Furthermore, the specimen still represents the largest known and youngest pterosaur from Australia. It had an estimated wingspan of between 3.6 and 4.9 metres, and came from the very youngest Cretaceous deposits, confirming that pterosaurs did indeed persist up to the cataclysmic Cretaceous/Tertiary boundary.

Another record of a neck vertebra similar to that of *Quetzalcoatlus* was reported from near Merigon in the French Pyrenees, and described in 1997 by a team of French palaeontologists headed by Eric Buffetaut at the University of Paris[12]. The specimen came from an animal with an estimated wingspan of close to 9 metres, showing that these gigantic pterosaurs were widespread across the world by the end of the Cretaceous, occurring in North America, the Middle East, Uzbekistan, Senegal, France and Australia. Why the reign of these fabulous beasts came to an abrupt end as suddenly as that of the dinosaurs may never be known. However, changing patterns of life in the seas brought about by rapidly changing sea levels and dramatic climatic shifts probably affected the food chain enough to disrupt their food supplies.

Perhaps some of the most exciting recent discoveries are those which give us valuable insights into the lifestyle of pterosaurs. Recently there has been an increase in the study of juvenile pterosaur growth. A pterosaur rookery has been discovered from an ancient floodplain in the Atacama Desert of South America, in sediments 110 million years old. In this deposit, Mike Bell of Cheltenham and Gloucester College in England, discovered thousands of bones of young pterosaurs in a 2 metre thick conglomerate. Analysis of the bone structure of these baby pterosaurs by Kevin Padian has shown that the bone formed early in development was packed with blood vessels, suggesting that the pterosaurs grew very rapidly after hatching. There is a lack of adult bones in this deposit implying that these young reptiles were incapable of flying. If the conglomerate of rock and bones was produced by a flash flood, the adults would have been able to fly away. Only if the juveniles had been capable of flight could they have escaped. The existence of their fossils suggests that they couldn't fly. As juveniles maybe they possessed extensive running and hunting abilities, or alternatively they were nurtured by their parents during the dangerous period when they were landbound.

Similar, highly vascularised bone structure has been recorded in other pterosaurs, including the Jurassic *Rhamphorhynchus* and the Late

Cretaceous *Pteranodon* by Chris Bennett. Bennett has shown that many of the specimens of *Rhamphorhynchus muensteri* he examined fell into three age classes[13]. These were not based on size, but on morphological characteristics. The youngest age class, consisting of immature juveniles, have a number of unfused skeletal elements, and the bone is incompletely ossified. Older juveniles developed more bone fusion. In the largest specimens, that probably had wingspans up to 1.8 metres, all the bones are ossified. In addition, the bones are covered by a hard outer layer, lacking blood vessels. This indicates that pterosaurs had determinate growth, and are likely to have stopped growth when they attained sexual maturity. Bennett interprets these three size classes as representing year-classes, each fossil sample representing a mass mortality. As such, Bennett considers that sexual maturity was reached after three years, at which time growth effectively ceased.

Bennett has also shown how rapidly the fourth finger that supported the wing must have grown during development in *Rhamphorhynchus.* In juveniles this wing finger is less than 10 times the humerus length, but increases to about 10 times humerus length in subadults. Like bats, this important finger would have undergone a great elongation after birth. Growth of such a long finger would have been very limited in the egg. This means that it is likely that newly hatched pterosaurs were incapable of flight. However, the faster this wing finger grew, the faster the youngster could take to the air and the greater its chance of survival.

To date some 60 different species of pterosaurs have been found, and due to the delicate nature of their bones and generally rare preservation, it is not unlikely that many more species existed, flourishing in diverse environments where their bones may not have been well preserved. The very sight of a 12 metre wingspan *Quetzalcoatlus* gliding gracefully over the ancient seas and scooping up fish with its metre-long beak would have been an awesome sight. Certainly no flying creatures since have ever approached the huge size of this animal. By the end of the Cretaceous the birds had proliferated and appear to have taken up the niche of insect eaters, a task requiring more flight manoeuvrability. We have no direct evidence of why the pterosaurs succumbed to extinction's fatal scythe, yet the birds survived the Cretaceous/Tertiary boundary mass extinction event almost unscathed. Perhaps it was the former's connection to the food chains of the marine realm that was their ultimate undoing as sea-level and climatic changes occurring over short time frames are likely to have wreaked havoc upon the aqueous ecosystems. These giant pterosaurs would have been little threat to most of the terrestrial animals of their day. It seems that the pterosaurs' long reign of terror and triumph in the ancient skies was inflicted only upon fishes.

12

Dinosaurs of the Twilight Zone

Polar dinosaur faunas

Standing high on top of an icy mountain about 900 km from the South Pole, I looked out over a vast sea of blue glacial ice, with rocky mountains pushing their way through the sky. I was there in late 1991, searching for ancient fish fossils. Only a year before a US team had pulled the partial skeleton of a giant meat-eating dinosaur from the top of Mt Kirkpatrick, about 200 km further south from where I stood. That dinosaur, since named *Cryolophosaurus*[1], lived about 190–200 million years ago, at a time when Antarctica was the forested central hub of the giant southern supercontinent Gondwana.

Antarctica today is the most remote and hostile place one can imagine, and such regions only support vertebrate life in or near their biomass-rich seas. Inland, there is little alive, and the thought of dinosaurs draped in heavy fur or featherdown coats roaming this extensive wasteland is somewhat ridiculous. Yet such a fanciful idea is not too far from the reconstructed scenario that palaeontologists have constructed for an area that was close to Antarctica 120 million years ago – southern Victoria in Australia. There, over the past decade, more than 10 000 bones, mostly of dinosaurs, have been recovered from a series of coastline outcrops of grey–blue sandstones. The dinosaurs are not particularly well preserved, except for some rare partially articulated skeletons. However, what makes these sites of world-wide interest is that they tell us how a community of dinosaurs and other animals and plants lived well within the Antarctic Circle, about 75° south, when southern Australia was welded to Antarctica.

The first dinosaur bone was found from the coastline of southern Australia around 1905 at Cape Patterson, eastern Victoria, by geologist William Ferguson, who was down there checking out the region for coal deposits[2]. The small, 4 cm long, sharp digit bone that he found has long been known as 'The Cape Patterson Claw' and was described as coming from a *Megalosaurus*-like beast – a large 'meat-eating' theropod dinosaur. In November 1978 a party consisting of Tim Flannery, now at the Australian Museum, Rob Glenie, a freelance geologist, and myself, decided to visit the site and have a go at finding more dinosaur bones. We didn't really expect to find anything, as the coastline had been searched by several fossil hunting groups over the years to no avail. It was a blustery day, light drizzle and strong winds lashing the barren coastline. Only minutes after scrambling down the rugged cliffs I picked up a rounded sandstone pebble. Running through it was the cross-section of a black dinosaur bone!

That day we found a few more bones, and thus began the great dinosaur search of the Cretaceous outcrops in southern Victoria. In the early 1980s, on the other side of the Victorian coast near Apollo Bay, a rich seam of dinosaur bones was discovered. The little bay housing this treasure trove was aptly named 'Dinosaur Cove'. Such was the richness of the find, that palaeontologists Tom Rich, of the Museum of Victoria, and Patricia Vickers-Rich, of Monash University, directed their field team to begin mining straight into the cliff, complete with tunnels and pit props. Shift-working volunteers laboured 24 hours around the clock because the expensive mining equipment was on short-term loan. This was the only way to achieve their goal – to extract all the bones from a single channel deposit. Many thousands of bones have now been discovered, and although only a small percentage of those have been prepared, it is clear that the fauna is surprisingly high in its diversity of small dinosaurs, mostly hypsilophodontids, that dominate the fauna[3]. In addition to dinosaurs, remains of plesiosaurs, pterosaurs, turtles, fish and a giant amphibian have been found[4].

Many different kinds of plants and insects occur, along with beautifully preserved fishes in an ancient lake depost near Koonwarra in Victoria, within the same sequence of rocks as those in which the dinosaur bones occur (the Strzlecki Group). When geologists first started looking at the environment that these animals and plants inhabited it became clear that the ground temperatures would have been very low. Oxygen-isotope analyses on the groundwater temperatures of the sediments yielded a mean average annual temperature of 0–5°C[5]. The southerly position of the sites indicate that for maybe three months of the year the region was in total darkness. These fossil deposits were laid down in the vast rift valleys that formed when Australia first began to be torn away from Antarctica, signalling the onset of the final breakup and death of the huge supercontinent Gondwana. Into these valleys poured huge volumes of volcanic-derived sediments, rapidly burying the animal and plant life in the channels of small braided streams and occasional lakes that pooled on the alluvial plains.

The whole scenario now becomes more interesting as one looks closely at the implications of the Victorian dinosaur fauna. For a start, most of the dinosaurs were small, and as the temperatures were very low, how did these creatures manage to stay alive? The old model of cold-blooded dinosaurs running around a freezing, dark landscape just doesn't hold water (or ice). Were they warm-blooded, as was suggested by Bob Bakker back in 1972 (and treated extensively in his *Dinosaur Heresies* book of 1986), and by Adrian Desmond in his *Hot-Blooded Dinosaurs* (1975)? Or did they use other more subtle strategies, like antifreeze enzymes, similar to those occurring in sub-Antarctic and Arctic fishes or in giant salamanders that inhabit near-frozen streams in the middle of the Japanese winter? Maybe a layer of fat protected them. Perhaps even a layer of fur or feathers? Did they huddle together to keep warm?

In the early editions of my book *Dinosaurs of Australia* I drew my *Allosaurus* with a layer of fine fur, but it just as well could have been of downy feathers. My rationale for this was that if pterosaurs, the flying reptiles that were close relatives of the dinosaurs, could have had furry wings and bodies, as has been documented from superbly preserved specimens from Kazakhstan, then why not dinosaurs? The presence of feathers on dinosaurs was suggested by Russian palaeontologist Sergei Kurzanov because of his small bird-like theropod *Avimimus*. The arm bones found associated with this dinosaur had grooves similar to those where feathers insert on modern birds. The presence of feathers on dinosaurs is not such a hard transition to accept. If chicken embryos are given retinoic acid (Vitamin A) at a certain stage of their development, the scale pattern on their legs is replaced by feathers, and the body which is normally covered by feathers, remains scaly.

Big trouble with little Chinese dinosaur! The recently discovered turkey-sized Sinosauropteryx, *a small dinosaur with feather-like fibres running down its back, from the Early Cretaceous of China, gives credence to the notion that other dinosaurs, living in polar climates such as in Victoria, may also have had insulating outer coverings of feathers or fur.*

About half a dozen fossil feathers have been found in the same Victorian sedimentary sequence as the dinosaurs are found in, at Koonwarra in Victoria[2]. These have usually been thought to represent the earliest birds in Australia. However, since recent studies are revealing that the overall physiology of dinosaurs was closer to an avian model than that of a basic reptile, like a lizard or crocodile, it is not surprising that some palaeoartists had predicted dinosaurs might well have had feathers. Possible evidence came in 1996 when the first discovery was made of a beautifully preserved dinosaur with a cover of what appeared to be small feathers[6]. This, and a subsequent second complete specimen from Liaoning, China, have been named *Sinosauropteryx*, and suggest that some dinosaurs living in the Early Cretaceous developed a cover of feathers. This was perhaps a necessary adaptation in the cold Antarctic forests of south-central Gondwana.

Recent research on superbly preserved *Sinosauropteryx* specimens[7] has revealed that while not being like feathers seen on living birds, these 'integumentary structures', as they are known, are similar to the plumules of modern birds, but lack the barbules and hooklets that characterise a true feather. Suggestions have been made that while being the precursors to feathers, these structures would not have enabled the dinosaur to fly, but may have been for display, or maybe for thermoregulation. As well as these structures, specimens of this little dinosaur have been found with their eyes preserved, with their stomach contents (a lizard in one case, a mammal in another), and in one specimen with a pair of eggs in the oviduct. Such amazingly preserved specimens will shed much light on both dinosaur and bird relationships, as well as provide insights into their life habits.

The great diversity of plants living in the forest with the polar dinosaurs of Victoria also supports the idea of a cold climate, but not the frozen snow-covered setting of present-day Antarctica. Today no trees grow in Antarctica. This is a relatively recent phenomenon, for only about 3 million years ago forests of southern beech (*Nothofagus*) were growing on mainland Antarctica, about 800 km from the South Pole, at the site of the Beardsmore Glacier[8]. The fossil leaves were found in glacial till deposits in 1993, and were dated as Pliocene, about 2 million years in age, on the basis of the associated spores and pollens in the same sediments. The great southern continent froze slowly. The thermal decline increased rapidly once the the circumpolar current enveloped the continent about 30 million years ago, quickly transforming the formerly lush continent into a polar desert.

The ancient Cretaceous forests of Victoria were dominated by large conifers, araucariacean pines, ginkgoes, ferns and seed-ferns, with the barest hint of the earliest flowering plants, angiosperms. In fact, one of the world's oldest fossil flowers (see Chapter 9) comes from the same deposits in eastern Victoria as the fossil feathers (the Koonwarra site), which also preserves a great diversity of insects, both as adults and larvae, along with some spiders and mites[9].

Yet the dinosaurs from Victoria that once roamed these dark, cold southern forests have raised more issues than the question of what physiological challenges the dinosaurs would have had to meet living in such harsh conditions. When we look closely at the types of dinosaurs and other animals that lived here, at the bottom of the Earth, 100 million years ago, we would expect to find that it fitted a model for the distribution of major dinosaur groups in time and space that is consistent with the total picture of what we know about dinosaurs as a whole. Nothing could be further from reality. On the contrary, the polar dinosaur fauna from Victoria is rewriting the story of dinosaur evolution. And it all started back in 1981, the day we found the *Allosaurus* bone.

My cousin Tim Flannery found the bone about 100 metres west of Eagles Nest, a prominent rock structure off the coast of Inverloch. After having been prepared out of the rock in the lab, eventually it was identified as an astragalus, or ankle bone, from a theropod dinosaur[10]. Surprisingly, if you don't have the skull, the astragalus is the second best bone to have if you want to name your dinosaur, as it is one of the most diagnostic bones in a theropod dinosaur skeleton. This is because the dinosaur walks with its weight transmitted down through the metatarsal bones that support the toes, similar to a bird. This is of course no mean coincidence, as birds are widely considered to be descended from bipedal theropod dinosaurs (see Chapter 13). The astragalus and calcaneum are the ankle and heel bones, respectively, which transmit all that mighty weight down to the toes, and thus have a complex and unusual structure, especially in more advanced theropods, in which the astragalus has a distinct protruberance of bone (called the dorsal process) that fits over an indentation on the end of the shin bone (tibia).

The theropods were predators represented by such well-known favourites as the mighty *Tyrannosaurus rex* and *Allosaurus*. The latter lived in North America some 140 million years ago in the Late Jurassic, so when our Victorian bone matched perfectly the shape of the astragalus of *Allosaurus*, and was identified as such, North American dinosaur experts pooh-poohed the idea[11]. How could an *Allosaurus*, a hunter that roamed the forests of North America, be found inhabiting polar forests some 40 million years later in the Antarctic? This was the first example of a chronological anomaly, or 'time slip' in the fauna, the first of many such surprises to come. The battle of words ensued in the scientific publications, with elegant titles that reinforced the Australian scientists' confidence in their identification such as 'Aussie *Allosaurus* after all'[12]. The conclusion that the bone came from an *Allosaurus* was reached after relentless comparisons with a wide range of theropod astragalus bones, searching the world comparing dinosaur foot bones for the perfect match, just like Cinderella trying on the glass slipper!

So, when such a time anomaly occurs, it is all too easy to dismiss it as a one-off event. It's only one bone; they've probably misidentified it; and anyhow, it doesn't mean much in the global scenario. But the next such anomaly came soon after. A large jaw had been found in the deposits at Dinosaur Cove, but unlike the search for *Allosaurus*'s slipper, identifying the jawbone was not so easy. It was compared with all major dinosaur jaws, but none were even vaguely similar. The mystery was solved in about 1990, when Anne Warren, from Latrobe

Isn't it amazing what you can do from one ankle bone? This big hairy ugly brute is a somewhat fanciful reconstruction of an Allosaurus *from the Early Cretaceous of Victoria, based entirely upon the astragalus, an ankle bone. From it, we know it is* Allosaurus*; from its size and robustness we can say it was a small, heavily built version; and from its geological context we can adorn it with insulation to protect it against the cold.*

University, suggested that maybe it wasn't dinosaur after all, but part of an amphibian jaw, from an extinct group called the labyrinthodonts.

This seemed absurd at the time because labyrinthodonts mostly became extinct at the end of the Triassic Period, some 200 million years ago, with just an odd straggler surviving into the Jurassic in China. In the late 1980s an amphibian was identified from the Early Jurassic of Queensland in Australia. This beast, named *Siderops*, meaning 'iron face', was the first such chronological anomaly in Australia suggesting the later survival of the short-faced labyrinthodonts known as brachyopoids. But to extend this group's range into the Early Cretaceous would push the age range of the group into even younger geological time – another 40–50 million years.

Soon more of the mystery bits from the Victorian sites were identified as amphibian remains, and now the tally includes perfectly preserved complete jaws, shoulder girdle bones, parts of the skull, and very recently most of an individual skull. It was a large creature. Some of the jaws are nearly a metre long, similar in size to those of *Siderops*. At the same time as making these finds, it was realised that there was an absence of crocodiles in the Victorian fauna. The connection was made by Tom Rich that perhaps these giant, crocodile-like amphibians had occupied the niche that crocodiles would later take over, but at this time, maybe it was simply just too cold for crocodiles, so close to the South Pole. Since the discovery of the Victorian Cretaceous labyrinthodonts, similar specimens from Russia have been identified. These, again, were not originally recognised as such because either the age of their site was not believed to be Cretaceous, or the bones were thought to be from animals other than labyrinthodonts.

The dinosaur faunas from Victoria are most unusual in that the large majority of bones come from the little hypsilophodont dinosaurs – the swift-footed little 'antelopes' of the dinosaur world[2]. Hypsilophodonts elsewhere were never the dominant element in any fauna, usually being just a minor component. The abundance of hypsilophodont bones and high diversity of types in the Victorian faunas is most unusual. Tom Rich has estimated that at least five or six different genera are found in the fauna[13], predominantly small animals a metre or less in size. Some of the best dinosaur fossils found during the decade of excavations include the skull of a little hypsilophodont, and two partial skeletons showing the articulated legs and part of the tail.

Opalised bones of hypsilophodonts from Lightning Ridge in New South Wales, that are of similar age to the Victorian bones, were described by the noted German palaeontologist Fredrich von Huene in 1932. Von Huene named a new dinosaur, *Fulgurotherium* (meaning 'lightning beast') based solely on the diagnostic features of the thigh bone, or femur[2]. Many years later this was proving to be a nightmare for Tom and Pat Rich, as they had many hundreds of hypsilophodont bones, mainly femurs, and so sorting out the different species naturally began with the diagnostic features of this one leg bone. The discoveries of a rare skull, along with jaws, teeth and partially articulated skeletons, then made diagnosis nigh impossible for the really nice specimens, as unless they had a femur preserved they were never sure which animal the bones came from. Eventually, through associated partial remains being found in the same beds, the skull was identified as being from the abundant species *Leaellynasaura amicagraphica*, named after the Rich's daughter Leaellyn, and in honour of their major sponsor, The National

Gazelles of the dinosaur world? Little hypsilophodont dinosaurs were the most abundant form living in the polar forest of southern Australia some 100 million years ago. They were swift-running plant-eaters that probably occupied a niche similar to that of small deer or gazelles today.

Geographic Society. Another sponsor, who donated the expensive earth-mining equipment was later honoured by having a new hypsilophodont dinosaur named after them – *Atlascopcosaurus* (no prizes for guessing who the sponsor was).

The delicate hypsilophodont skull found at Dinosaur Cove was only about the size of a chicken skull, yet it proved to be most interesting in that the rear part of the skull roof was missing, exposing the sedimentary rock that infilled the brain cavity (an endocranial cast). The skull itself has very large orbital notches for the eyes, and the endocranial cast shows that the dinosaur had very large optic lobes, suggesting that this little dinosaur had large eyes and excellent vision. This may possibly have been an adaptation for the long periods of darkness, maybe up to three months a year, experienced by these dinosaurs in the chilly polar forests[3]. Furthermore, some of the hypsilophodont dinosaur leg bones show abnormalities. One femur shows a large degree of additional bone growth, indicating osteomyelitis had set in possibly from infection developing after a wound was inflicted[13]. This suggests that the forests may not have been too heavily populated by predators, allowing small, wounded herbivores a second chance at life.

Then in the early 1990s a most remarkable dinosaur discovery was made at Dinosaur Cove. A long leg bone, specifically a femur about the same size as a human femur, along with a smaller femur, were also identified from the same deposits. The thin, slender shaft of the bones have well-preserved ends, providing enough morphological information not only to identify the dinosaur family of the bones' owner, but also diagnostic enough to erect a new genus of dinosaur. The bones came from one of the fleet-footed, ostrich-like dinosaurs, ornithomimosaurs. This group was thought to have lived only in North America and Asia, and predominantly in the Late Cretaceous, although an early, ancestral form is known from the Early Cretaceous of Mongolia. Still, the occurrence of a definite member of the ornithomimosaurid family in the Early Cretaceous of Australia, near the South Pole, was completely unexpected, and suggested an early origin for the group in Gondwana.

The Australian ornithomimosaur was named *Timimus*, meaning 'Tim's mimic' in honour of both Tim Rich, the son of Tom and Pat Rich, and Tim Flannery, who found many of the early dinosaur bones from the Victorian sites[14]. *Timimus* grew to further fame during the heights of dinosaur mania in Australia soon after the release of Spielberg's *Jurassic Park* in 1993, when it adorned an Australian 45c postage stamp. Based on the size and proportions of the thigh bone, this colourful reconstruction of *Timimus* was sent zooming all around the country and dispatched to the far reaches of the globe, at a much faster pace than its real-life counterparts could have ever dreamed of!

The well-established foundation of dinosaur origins and distribution was beginning to develop cracks in the early 1990s, when still more devastating discoveries came from Victoria. This time, an ulna of a little quadrupedal dinosaur was found. The bone was relatively complete, but missing one small end. Nonetheless it was dutifully carried around the world by Tom and Pat Rich in the search for a match with one from a well-known family of dinosaurs. Eventually the bone was matched perfectly with that of a ceratopsian[13], the horned dinosaur family that contains such favourites as *Triceratops*. When it was shown to North American dinosaur expert Dale Russell, of the National Museum of Natural Sciences in Ottawa, Canada, lying side by side with its doppleganger, from the North American dinosaur *Leptoceratops*, he exclaimed 'we are in violent agreement – that sure looks like a ceratopsian!'. What was particularly surprising was that the ceratopsians are a group known to have an excellent fossil record, spanning a precise range of time and space – all were from North America or Asia (Laurasia in the Cretaceous), and all were limited to the Late Cretaceous age.

There are only two logical possible explanations for the Victorian bone, representing a beast that had lived some 20 million years before the first (and previously only) ceratopsians had appeared in the

Northern Hemisphere. The first, and most plausible, is that the ceratopsian dinosaurs first evolved in Gondwana in the Early Cretaceous, then migrated northwards before the breakup of Gondwana, eventually radiating and becoming established in the Laurasian continent in the Late Cretaceous. This hypothesis implies that other primitive ceratopsians should be found in the intervening areas between the Early and Late Cretaceous, and that their absence is most likely an artefact of the incompleteness of the fossil record. It follows the simple rule that the fossil record preserves only those creatures lucky enough to die in an actively accumulating sedimentary basin that will later survive tectonic upheavals and one day expose its treasures to the surface.

The second hypothesis requires more imagination, but in scientific terms it is equally likely based on the evidence at hand (one ulna). It is that the Victorian 'ceratopsian' actually belongs to a parallel group of dinosaurs that evolved independently of the well-known ceratopsians in North America and Asia, and is thus of similar form and skeletal shape, yet not closely related. Such parallel evolution often occurs, and many modern examples illustrate the point. For example, the thylacine, or Tasmanian tiger (*Thylacinus cynocephalus*), is a very dog-like marsupial that evolved within Australia, and has no close relationship with dogs, wolves or foxes from Africa or Europe. Indeed, the thylacine is probably more closely related to humans, in terms of the early time of divergence of the main lineages (marsupials, primates), than it is to the line leading to dogs (Carnivora). What we need is more skeletal remains of this enigmatic Victorian 'ceratopsian' in order to solve the mystery. Either way it is scientifically interesting, as it either pinpoints the earliest time and place for the first ceratopsians, and indicates their later migration, or, it indicates the existence of a completely new undiscovered major group of dinosaurs!

Further pieces of evidence for the polar regions of Gondwana having been home to many of the first groups of dinosaurs fell into place in the mid-1990s, when isolated mystery bones from small theropod dinosaurs were identifed as being from possible oviraptosaurids[14]. These dinosaurs were inappropriately named 'egg stealers', as early skeletons found in Mongolia of *Oviraptor* (meaning 'egg thief') were found near fossilised nests of what were thought to be eggs of a little ceratopsian called *Protoceratops*. Only in the 1990s did expeditions from the American Museum of Natural History led by Mark Norell find actual *Oviraptor* fossils sitting astride nests of their own eggs! They were not stealing the eggs, but probably brooding them. The Victorian dinosaur fauna now contained another example of a well-known Late Cretaceous North American–Asian group of dinosaurs, further supporting the overall story. Most recently dromaeosaurid dinosaurs have been identified from Victoria (and possibly Lightning Ridge, NSW). These include the well-known 'raptors' of the movie 'Jurassic Park', which were actually more like oversized *Deinonychus antirrophus* rather than the sleek *Velociraptor*. This group does extend back to the Early Cretaceous, but again lacked a fossil record from Gondwana.

The Victorian dinosaurs are not the only polar dinosaurs known. Dinosaur discoveries from Arctic Canada, Alaska, Siberia and Spitzbergen, all within or near the Arctic Circle, suggest that the Northern Hemisphere had its thriving cold-climate dinosaur faunas as well. Tom Rich has been quick to point out that not as much time and effort has been spent collecting at many of the Northern Hemisphere polar dinosaur sites compared with the Victorian sites[15]. Consequently, attempts at comparing faunas on either side of the world are not yet meaningful, especially as the majority of the Arctic dinosaur faunas are younger, being of Late Cretaceous age. Still, further analysis of these sites could reveal reasons why the Victorian dinosaur faunas had a high dominance of groups like the hypsilophodontids, and whether certain dinosaur families were more temperature tolerant than others.

The overall picture that is emerging of the southern polar dinosaurs, based on several sites, such as Victoria, New Zealand and Antarctica, is of abundant small plant-eating ornithopods (hypsilophodontids and dryosaurs), small to medium theropods (allosaurs, ornithomimosaurs, oviraptorosaurs, dromaeosaurs), rare quadrupedal plant-eaters (ankylosaurs, ceratopsians), and a virtual absence of the giant sauropods, like *Brachiosaurus* or *Diplodocus*. Perhaps these monsters could only thrive in areas of high plant biomass, and where easily accessible and digestible species dominated the forests? In addition to these dinosaurs and giant amphibians, fragmentary remains indicate the presence of turtles, plesiosaurians, and flying reptiles, or pterosaurs.

The whole biota of these fossil deposits has now been studied in detail, bringing evidence from the animals, the physiological adaptations of the plants, the insects, even the fishes, together with geological observations of the sedimentary rocks, their diagenesis, isotopic studies of temperatures and so on. Despite this, many unsolved mysteries remain which can only be clarified with better discoveries from future digs. The secrets of the Victorian polar dinosaurs and their cold, dark world will still be kept for some years to come, as although many thousands of bones have been found and few studied, the vast majority remain unprepared, and unstudied.

So, eagerly each year we await more startling discoveries from southern Victoria's extensive coastline of exposed Cretaceous sediments. Such finds do more than just fill in the picture of what life was like at the bottom of the Earth 100 million years ago; they also suggest that the origins of some of the major dinosaur groups could have been born out of such a harsh environment. The dinosaurs did migrate as far as the land bridges of the day would have allowed them, which was anywhere in the world before the end of the Late Jurassic. The continental positions suggest that the southern polar dinosaur fauna comprises mostly a relict assemblage of groups that evolved there or in nearby Gondwana regions, such as Antarctica, South America or southern Africa, and later moved out north into North America and Asia before the incipient breakup of the supercontinent Pangaea. They only had one direction to go from their home base near the South Pole. North to Alaska, north the rush is on!

13

Birds of a Feather Fossilise Together

The early evolution of birds

The next time that you saunter over to your bird cage to give your budgie some seed, spare a thought for the plight of your feathered buddy. I don't just mean the fact that its immediate wild ancestor had at least half of Australia to fly around in, compared with the possibly less than palatial cage in which the bird now perches. I'm thinking of even more distant ancestors of this budgie: birds who were just a stone's throw away from being a dinosaur. For what was played out on the evolutionary stage some 150 million years ago or so, was the flight to freedom of one small group of dinosaurs – literally the evolution of flight. Given half a chance wouldn't your budgie emulate its distant ancestor and take its own flight to freedom?

Since the discovery in the nineteenth century of a number of spectacular fossils from lithographic limestones in Solnhofen in Germany, the early bird *Archaeopteryx* has stood out like a shining light in our understanding of the evolution of birds, principally because that was all the fossil evidence we had. But in the last couple of years a revolution has come to avian palaeontology, in particular a revolution in how we interpret the early evolution of birds. This has been fuelled by the discovery of a number of spectacular fossils. Barely a month seems to go past without yet another amazing fossil bird specimen turning up in China or Spain or Madagascar, or wherever. As a consequence, our understanding of the evolution of birds, and of avian flight, is changing all the time.

Yet despite these many new discoveries, none has so far pre-dated *Archaeopteryx* – at least, none that have been substantiated. One Late Triassic fossil, called *Protoavis*, was considered to be an even earlier bird than *Archaeopteryx* by the palaeontologist who described it, Sankar Chatterjee of the University of Texas at Austin. Chatterjee saw this creature as being closer to modern birds than *Archaeopteryx*. However, the general consensus of opinion amongst bird palaeontologists is that *Protoavis* is not their pigeon. The specimen consists of many incomplete, fragmentary bones, and is has been suggested by some scientists that the different bones may belong to more than one species. Another suggestion that birds may have once existed in the Triassic came in the 1970s, when fossil footprints from Late Triassic to Early Jurassic rocks in South Africa were described as having been made by birds. However, the close similarity of the feet of early birds to those of small dinosaurs makes assigning such footprints to birds questionable.

Although, as I shall discuss, new finds from China may be close in age to *Archaeopteryx*, it is still the fossils from Germany, many collected so long ago, that form the benchmark against which all other finds are assessed. Even so, many aspects of the biology of *Archaeopteryx*, in particular its flying ability, have been the subject of much debate. For instance, some have argued strongly that *Archaeopteryx* was a tree-dwelling bird, whereas others have argued just as vehemently that it shows no adaptations whatsoever for living in trees. One of the problems is that we still do not have detailed information on some aspects of its anatomy that are crucial to a fuller understanding of the biology of this bird. In particular, little is known about its braincase, its orbit and palate. A few scientists are firmly of the opinion that *Archaeopteryx* was predominantly a land-dwelling animal, with wings that had little power of flight. But most ascribe some sort of flying ability to *Archaeopteryx*, believing that it may even have approached the flying ability of some living birds.

In 1995, when reviewing the arguments for and against the flying proficiency of *Archaeopteryx*, Luis Chiappe of the American Museum of Natural History, New York City, pointed out that there is a fundamental problem with reconstructions that have *Archaeopteryx* perched on a branch up a tree like a smug pigeon[1]. Palaeoenvironmental reconstructions of the land around the lagoon into which the *Archaeopteryx* individuals plummeted and were ultimately fossilised indicate a distinct lack of trees. Rather than being flanked by woods or forests, only small plants, less than 3 metres high, surrounded the lake[2]. There is little doubt that there were major morphological differences between *Archaeopteryx* and modern birds (not only in its possession of a beak full of teeth, but also its longer tail that is not fan-shaped, absence of alula feathers and no carinate sternum), implying that however it lived, it was probably unlike any living bird.

However, detailed analyses of the anatomy of *Archaeopteryx* reveal a lot about how this bird lived. Tony Thulborn and Tim Hamley of the University of Queensland have argued that *Archaeopteryx* was an agile hunter that lived along the edges of lakes[3]. The structures of the hindlimbs, the teeth and jaws suggest that *Archaeopteryx* was not an insect-eater that lived up in trees, but an active, ground-dwelling bird that hunted for its prey along the shores of lakes. Thulborn and Hamley suggest that rather than being used exclusively for flight the wings could have been used in much the same way that herons and egrets use their wings today: to make an arched canopy while foraging in shallow water for worms or fishes. The reliance on water has led to the suggestion that perhaps the ability of birds to fly first arose from these water-inhabiting dinosaur descendants. Their simple, primitive wings may perhaps have functioned well enough to allow them short flights from wave-crest to wave-crest.

Just where birds came from has engendered almost as much debate as to what killed off the dinosaurs. Opinions have essentially polarised into those who are convinced that birds evolved from theropod dinosaurs and those who consider that they evolved from a small arboreal group of basal archosaurs of thecodont grade, some time in the Triassic. Johann Welman of the National Museum of South Africa in Bloemfontein, for instance, has suggested that birds evolved from an archosaur like *Euparkeria*[4]. The scientists who take this point of view would counter any argument about the amazingly close anatomical similarity between theropod dinosaurs and early birds as being merely examples of convergent evolution. In their model flight evolved from 'the tree down', with the earliest birds having evolved from one of these arboreal archosaurs via a gliding intermediary. However, the great explosion in discoveries of Cretaceous birds made in the last few years, plus the discovery of some very bird-like theropod dinoaurs, along with reinterpretations of the mode of life of *Archaeopteryx*, all suggest that the evolution of flight was more likely 'from the ground up', and from theropod dinosaurs.

Is this a bird or not? In many ways Archaeopteryx *was just another small feathered dinosaur. It lived in the Late Jurassic of Germany about 150 million years ago, had wings, feathers and could probably fly well. Its skeleton, however, is indistinguishable from other small predatory dinosaurs of its time, even with its long arms and extended fingers. Well, all right, call it a bird if you want to, it really makes little difference one way or another!*

Which particular group of theropod dinosaurs could *Archaeopteryx* possibly have evolved from? Most favoured in recent years has been the Dromaeosauridae, in particular forms such as *Deinonychus*. There have been those who have argued against a theropod origin for birds. This has centred, to a large degree, on the premise that *Archaeopteryx* predates likely theropod ancestors. However, many non-avian dinosaurs that are close to the origin of birds lived in the Early Cretaceous, such as *Utahraptor*, *Deinonychus* and *Sinornithoides*. The remains of *Deinonychus* have even been found in the Late Jurassic Morrison Formation in North America.

The recent description by Fernando Novas and Pablo Puerta from Argentina of a very bird-like theropod from Cretaceous rocks in Patagonia that they have called *Unenlagia*, sheds yet more light on the dinosaur-bird transition[5]. Taking its name from the Mapuche Indian words *uñen* and *lag*, meaning, respectively 'half' and 'bird', this 2 metre long theropod dinosaur, despite being younger than *Archaeopteryx*, is probably the most bird-like theropod found so far. The shape of its scapula, pelvis and hindlimb are very similar to the same structures in *Archaeopteryx*, but other features are much more theropod-like. Although *Unenlagia* has enough features to characterise it as a dinosaur rather than a bird, the structure of its forelimbs indicates that it could fold them in a very bird-like manner. Its possession of extensive forelimb elevation indicates that this important prerequisite for powered, flapping flight was already present in ground-dwelling theropod dinosaurs, even though they lacked the ability to fly. Such capacity for producing a pronounced upstroke was a critical precondition for flight, to enable winged theropods to leave the ground. Moreover, this new evidence would suggest that powered flight originated prior to the ability to glide.

Unenlagia is therefore morphologically intermediate between dromaeosaurid dinosaurs, like *Deinonychus*, and *Archaeopteryx*, making it the most bird-like dinosaur known. However, its large size and shorter arms than *Archaeopteryx* indicate that it was not capable of flight. However, as Novas and Puerta have pointed out, there were no structural differences between the the bones of the forelimb of *Archaeopteryx* and of *Unenlagia*: they differ only in proportions. For flight to be acquired all that was then necessary was a reduction in body mass, an increase in forelimb size and, of course, the enlargement of feathers.

The attainment of a number of features characteristic of early birds has arisen, Tony Thulborn has argued, by the acquisition of what are essentially paedomorphic features – ancestral juvenile features retained by descendent adults, in other words, traits possessed by juvenile dinosaurs. In the mid-1980s he suggested that juvenile theropods may have had a covering of feathers that were used as an insulating blanket[6]. These structures were subsequently retained by adults of the early birds. Others have likewise argued that the original function of feathers was not to aid flight, but was to assist in thermoregulation, either to conserve body heat or to protect from solar radiation. While this idea gained a limited degree of acceptance, it was given a tremendous boost in 1996 with one of the more astonishing fossil finds of the century. In the Yixian Formation in Liaoning Province in China, specimens of small, coelurosaurian-like dinosaurs were uncovered, complete with manes of feather-like fibres running down their backs. Called *Sinosauropteryx*, these specimens are probably Early Cretaceous (about 125 million years) in age (see Chapter 12).

Just what is it that characterises the earliest of birds? Compared with their presumed dinosaurian ancestors they have a small body size and unfused bones plus some skull characteristics found in juvenile dinosaurs, such as inflated brain case, large orbit and tooth shape. A recent discovery of a nest of dinosaur eggs in the Gobi Desert in Mongolia that also contained two embryonic dromaeosaurid skulls thought to belong to *Velociraptor*, shows just how similar the shape of the teeth of early birds was to that of baby dinosaurs. In these embryonic dinosaurs the teeth are simple, peg-like structures, unlike the more complex adult teeth, but very similar to the teeth present in early, primitive adult birds[7].

In contrast to these ancestral juvenile features, *Archaeopteryx* is characterised by its relatively much larger forelimbs (that became the wings), the opposite situation to that found in dinosaurs. In particular, the manus was enormous, being about 40% of the total length of the forelimb. Thulborn and Hamley are of the opinion that it was so long that its clawed digits would have dragged along the ground as it walked along[3]. While the claw is often seen as an adaptation to help the bird climb, it is similar to the clawed digits of theropods, that almost certainly kept their feet on the ground. Most probably, as with dinosaurs, the claws were used to grab prey, before eating it. The general view seems to be that the enlargement of the forelimb did not take place as a means of enabling flight to occur. If it was, then the manus would not have been so big. More likely its enlargement was most closely related to its adaptation to assist in predation. Only later did it become secondarily adapted for use as a means of locomotion.

One way in which *Archaeopteryx* differs quite a lot from similar theropod dinosaurs is in its short tail. Along with this shortening of the tail there was a change in the shape of the pelvis that resulted in the centre of gravity being shifted further forward than in a normal theropod. Thus the counter-balancing role of the tail was reduced. This inherent instability would have enabled it to make quick, forward lunges in pursuit of its prey, forever poised like the hunched villain of Victorian melodramas, ever ready to pounce on its poor, unsuspecting victim.

Archaeopteryx appears to have had less skeletal fusion than living birds. Interestingly, the pelvis of the 'half-bird', *Unenlagia* is not fused. As Larry Martin of the University of Kansas has pointed out, a list of the unfused or poorly fused elements of *Archaeopteryx* is much like a catalogue of the juvenile condition of living birds[8]. These bones are also unfused in non-avian dinosaurs, as well as in some other early birds, such as *Iberomesornis*. Martin has suggested that the way modern birds change as they grow up can shed light on the origins of birds. This is because many characteristics, such as pubic reflexion, prolongation of the ilium, and increased fusion, are first seen in the fossil record in a similar sequence to their appearance during development. For instance, adult birds that lived during Cretaceous times show some features like those that occur early in the development of living birds, in particular retention of distinct sutures in the skull. However, in most living birds the sutures are lost in the adults, although there are a few exceptions, such as *Tinamous* which retains open fronto-parietal sutures in the adults. Like the young of living birds, the sternum and parts of the ribs of *Archaeopteryx* generally are not ossified.

The great proliferation in fossil bird discoveries from Cretaceous rocks over the last few years makes it apparent that, like the explosive radiation in flowering plants and insects, birds also underwent a great evolutionary radiation at much the same time in the Early Cretaceous. Whether the bird radiation was related to these other two events has not really been addressed. However, it seems possible that with the geologically rapid appearance of a sumptuous array of fruits and flowers during the Early Cretaceous, and the accompanying larger feast of insects, birds would have benefited greatly. As new food resources became available, selection would surely have favoured a greater range of animals that could feed on them. And when it came to the top of the arboreal food chain, birds seem to have taken over from the pterosaurs that went into evolutionary decline during the Cretaceous. Pterosaurs of this age were larger than their Jurassic precursors. Perhaps they were too large to take advantage of the new flower, fruit and insect-eating niches that opened up.

The most important recent finds of bird fossils, in terms of what they can tell us about early bird evolution, have been made in Spain, China and Mongolia. A number of these can be placed into a group of birds known as the Enantiornithes, a diverse group of flying birds whose fossilised remains have been found on six continents. Some were tiny and toothed, like *Cathayornis* and *Sinornis*, while *Enantiornis* had a wingspan of 1 metre. Others, like *Gobipteryx* had a horny bill without teeth[1]. *Cathayornis* possessed a skull very much like that of *Archaeopteryx*[9]. The Enantiornithes has unique bone patterns in the feet and wings, unlike all other birds, and in many ways is intermediate between *Archaeopteryx* and the modern groups of birds. Kevin Padian of the Museum of Paleontology at the University of California in Berkeley has described enantiornithines as the Cretaceous birds that 'had strange shoulder girdles [and] warped foot bones'[10].

Luis Chiappe suggests that the intermediate morphology of enantiornithines reflects also an intermediate position in terms of its evolutionary history[1]. This idea has been bolstered by the recent discovery of a nesting bird from Early Cretaceous limestones in the Spanish Serra de El Montsec in the south-central Pyrenees. This exquisitely preserved immature nestling shows a mixture of both primitive and advanced avian characteristics[11]. The head is very primitive. Like *Archaeopteryx* it has a toothed skull, large orbit and vaulted braincase. Another feature of the skull that links it with non-avian dinosaurs is the presence of a bone behind the eye called the postorbital bone. This is not present in modern birds. The nestling was also similar to *Archaeopteryx* and many non-avian dinosaurs in having fewer neck vertebrae than living birds, just nine. The wing, however, was much more advanced, with wing proportions like those of modern birds indicating that it was a much more competent flyer than *Archaeopteryx*. In bird evolution the wings seems to have led the way.

There are other birds, however, that formed part of this Cretaceous evolutionary radiation, which cannot be neatly pigeon-holed (if you will excuse the avian pun). Some of these early birds were very small. For instance, *Noguerornis*, from very Early Cretaceous limestone in northern Spain, was about the size of a finch. Yet it tells an important story about the evolution of flight because it possessed certain features that indicate that this bird had improved flying ability[1]. For instance, individual manual elements were fused into a more rigid structure and the distal part of the wing was longer than in *Archaeopteryx*. Another tiny sparrow-sized bird from slightly younger Spanish rocks, called *Iberomesornis*, was probably an even better flier, as indicated by its possession of an ulna that is longer than its humerus. Perhaps of even more significance is that the anatomy of the foot shows that this bird was capable of perching, indicating that true arboreality evolved very early in the evolution of birds.

Another, more recently described enantiornithine from the Early Cretaceous of Spain has shed even more light on the evolution of flight, as well as revealing details about what some of these early birds fed upon[12]. This small fossil bird, named *Eoalulavis*, was about the size of a goldfinch (with a wingspan of about 17 cm) and was found in the Late Barremian (about 115 million years old) strata at Las Hoyas in the province of Cuenca in Spain. Because of the excellent quality of preservation of this specimen, the feathers show exquisite detail. What is especially significant about the *Eoalulavis* specimen is that the feathers

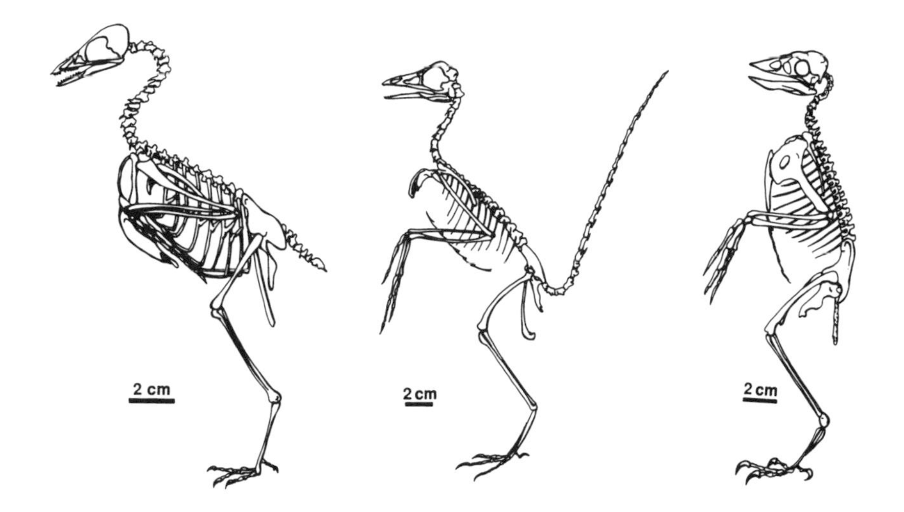

From dinosaur to bird in three easy skeletal steps. On the left is the feathered dinosaur/bird Archaeopteryx; *centre and right are two Early Cretaceous birds from China;* Chaoyangia, *still retaining teeth (centre), and* Liaoningornis, *without teeth (right), the latter also being the first bird to have a keeled sternum.* Archaeopteryx *was about crow sized; the other two about the size of pigeons. Based on drawings in reference 13.*

are still in natural life position. Not only are the primary feathers still connected to the phalanges and metacarpals, and the secondary feathers attached to the ulna, but the alula is still in its natural position. Also known as the 'bastard wing', these feathers are attached to the digits and perform an important role in helping to reduce stalling in flying birds, by allowing low-speed flight and manoeuvrability. They are particularly important in this regard in birds-of-prey. For birds to maintain lift while reducing speed they must increase the angle of attack of the wing. The alula, which is situated in the middle of the leading edge of the wing, allows the wing to create lift and reduces turbulence. It acts much like leading-edge flaps on the wings of an aeroplane. It automatically lifts when pressure above the wing is reduced. The alula usually only comes into play in modern birds when they take off or land, or, as in birds-of-prey, when they fly very slowly.

The alula in *Eoalulavis* is the oldest known example of these feathers and shows that such sophisticated structures had already evolved by the Early Cretaceous. Interestingly, it is not present in *Archaeopteryx*, adding further weight to the arguments that *Archaeopteryx* was not very accomplished at flying. Along with the alula,

Eoalulavis possessed quite advanced skeletal structures associated with accomplished flight, involving the ability to fly at low speeds and to undertake quite complex manoeuvres.

The exquisite preservation of Las Hoyas material is further shown by the existence of the gut contents of the bird, in which its last meal is preserved. What this reveals is that *Eoalulavis* fed on crustaceans. This implies that the bird probably had aquatic feeding habits.

Like these Spanish birds, the Chinese enantiornithines were also small and more than 30 individuals have been found at seven localities. The oldest have been found in the lake deposits in Liaoning Province that are about 120 million years old, although this age has been the subject of some debate. Both *Sinornis* and *Cathayornis* are enantiornithines. Indeed, enantiornithines appear to have been the dominant birds of Cretaceous ecosystems[1]. Like the Spanish birds they appear to have had the ability to perch. Arguably the most significant finds have been of a number of complete specimens of the 'wise old bird', *Confuciusornis.* Although some have argued that *Confuciusornis* was an enantiornithine, Luis Chiappe is of the opinion that it is far more primitive than other Chinese birds in some characteristics and may therefore not be an enantiornithine.

The original description of *Confuciusornis* pointed out that similarities in leg and pelvic structure to *Archaeopteryx* imply that *Confuciusornis* may also have had an upright posture[9]. Moreover, it had a similar wing structure, with long digits terminated by claws. The nature of the foot bones, with strongly recurved claws, was taken to imply that it was arboreal. Yet there is some debate over whether the structure of the bones of the feet would actually have enabled it to perch. *Confuciusornis* is an odd mixture of very primitive and advanced anatomical features[1]. One of these more advanced characteristics is the absence of teeth in its horny beak, many of its contemporaries having had beakfulls of teeth. Many of the specimens that have been discovered have their feathers preserved. These show that the bird had a long, feathered tail, not unlike that of a magpie, and a well-developed wing. The disposition of the feathers indicates that it was a capable flier. The presence of many specimens in the Yixian Formation prompts the question of whether these birds lived in relatively complex communal systems.

The same deposits have yielded another bird, *Liaoningornis.* Unlike some of the Early Cretaceous birds, this one possessed a keeled sternum. A number of other anatomical features indicate that this was one of the earliest ornithurine birds. Modern researchers consider that birds, the class Aves, are divisible into two groups, the Sauriurae, containing *Archaeopteryx*, *Confuciusornis* and Enantiornithes, and the Ornithurae, which contains all other known birds. It also possessed sharply recurved claws, indicating that it was largely arboreal. A distinctive difference between enantiornithine and ornithurine birds is the possession by ornithurines of a long posterior abdominal extension of the sternum. The effect of this was to provide ornithurine birds with a capacity for higher rates of oxygen consumption as they flew, similar to the method of air-sac breathing of modern birds[1]. Another fundamental difference between enantiornithines and more modern birds is their apparent lack of endothermy. But more on that shortly.

Not a great deal is known about the early ornithurine birds, as much of the early fossil material is very fragmentary. Other early forms include *Ambiortus* from the Early Cretaceous of Mongolia and *Gansus* and *Chaoyangia*, both from China. *Chaoyangia* is the most common fossil bird in the Chinese deposits, more than 25 specimens having been located. One of these shows that these birds possessed teeth. Reconstructions suggest that this bird was a wader, with small claws on its feet, a large rib cage and a shoulder girdle that is surprisingly modern in appearance[13]. *Ambiortus*, unlike modern birds, has two well-developed wing phalanges, plus a claw on its second digit. Morphologically this bird, and other early ornithurines, are quite distinct from the enantiornithines. Indeed, in many aspects of their anatomy, in particular their pelvis and hindlimb, Early Cretaceous ornithurines are very modern in appearance[1]. However, the acquisition of these more modern features came after the evolution of methods of flying comparable with living birds, which appear to have been achieved very early in bird evolution.

More is known about the Late Cretaceous ornithurines, in particular the Hesperornithiformes. These were flightless, diving birds that possessed teeth and probably propelled themselves through the water using their feet. Although first appearing in the Early Cretaceous (Aptian), most fossils come from younger strata. Fossils of these birds have been found in the western interior of North America and in the Turgai Strait, at the region between eastern Europe and western Asia which linked the ancient Tethys Sea with the Arctic Ocean. In addition to being found in marine deposits their remains have also been found in continental and lake deposits, indicating that they were not only geographically, but also ecologically, widespread. Although some studies

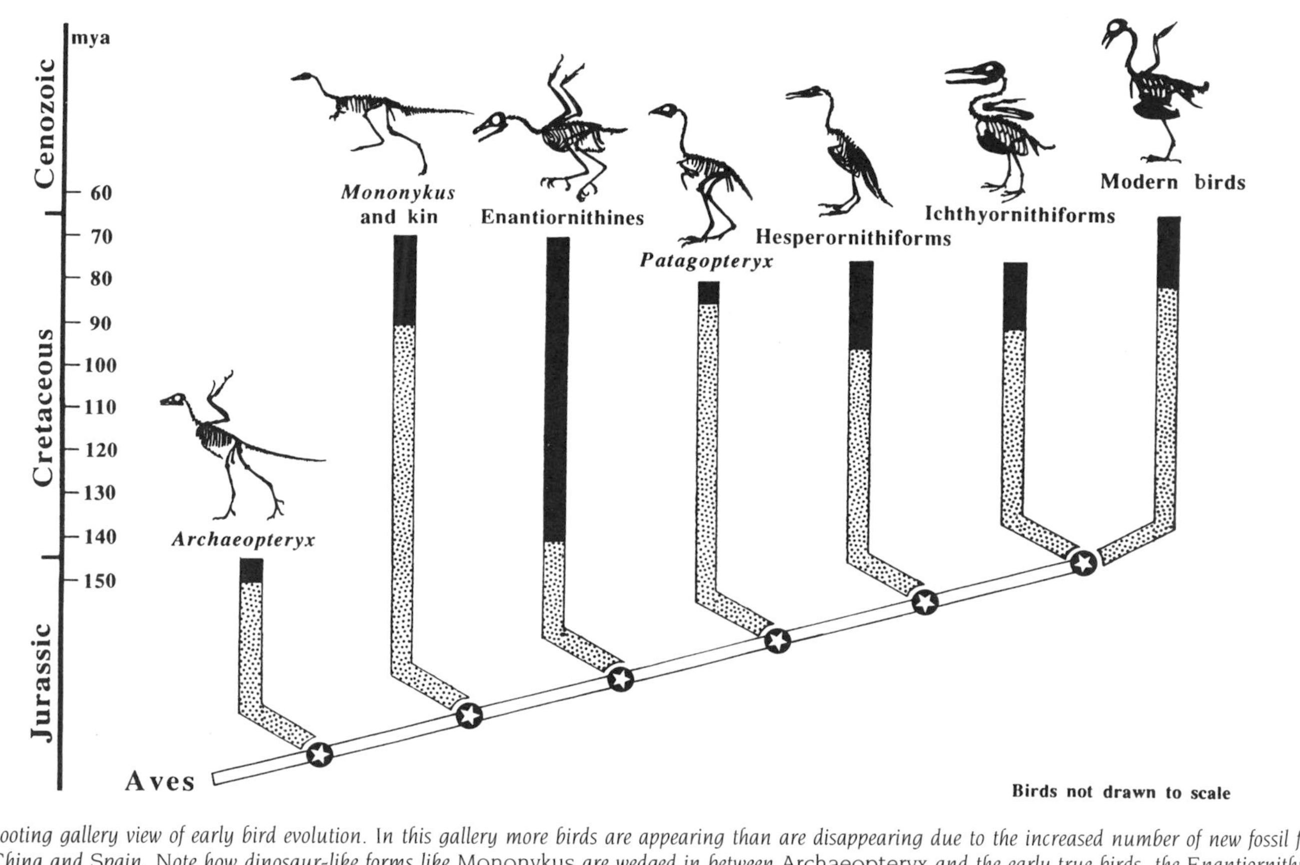

The shooting gallery view of early bird evolution. In this gallery more birds are appearing than are disappearing due to the increased number of new fossil finds from China and Spain. Note how dinosaur-like forms like Mononykus *are wedged in between* Archaeopteryx *and the early true birds, the* Enantiornithines. *The black line represents the actual fossil record known for each lineage.* Based on information and illustrations in reference 1.

have suggested that they might be primitive grebes or loons, the general consensus is that they are early ornithurines that became secondarily flightless.

By the end of the early Cretaceous, about 100 million years ago, there was a greater diversity in the size, as well as lifestyles of birds. There were some flightless birds, foot-propelled divers and waders, as well as tree-dwellers[1]. Amongst the more amazing flightless birds are *Mononykus* from central Asia and the hen-sized *Patagopteryx* from Patagonia. *Patagopteryx* seems to have evolved flightlessness independently of all other birds, and is unique in having some bones of the upper jaw fused to the palate[14]. *Mononykus* was a most peculiar bird. This flightless, stout-limbed, turkey-sized bird had close relatives living in Patagonia (*Alvarezsaurus* and *Patagonykus*). These creatures were so weird that there has been a good deal of controversy over whether they were really birds or non-avian dinosaurs. Apart from having a very greatly reduced ulna and radius, and a single large digit, *Mononykus* is very bird-like in a number of features, in particular its skull[15]. However, like *Archaeopteryx* its hindlimb functioned very much like that of theropod dinosaurs. What has concerned a number of palaeontologists is the fact that in most other respects *Mononykus* is just so different from *Archaeopteryx*. However, what they seem to forget is that probably these two forms were separated in time by well over 60 million years. That is a long time in evolution.

One of the more exciting recent advances in palaeontology is the analysis of fossil bone microstructure to unravel details about the life histories of long extinct animals. Bone microstructures can reveal much about growth rates. Analysis of the Cretaceous bird *Hesperornis* has revealed that its bone was laid down like that of modern birds, rapidly and uninterruptedly. Modern birds grow rapidly early in development, reaching maturity and final adult size usually within a year. In contrast, studies of bone microstructure of enantiornithines reveals a very different pattern, suggesting a quite different method of growth. These birds appear not to have grown continuously, but periodically. The bones reveal periods of growth interspersed with interruptions in bone deposition. Moreover, rather than having well-vascularised bone, that had lots of blood vessels running through it (necessary for rapid growth), enantiornithine bone was lamellar and poorly vascularised, implying slow rates of deposition.

Similar periodicity in bone growth in reptiles is due to annual periods of growth and non-growth. Such may have been the case in enantiornithines. In those that have been studied, four or five lines of arrested growth have been identified, indicating four to five years of post-hatching growth. Modern birds show a range of developmental patterns, from hatching as naked and blind altricial forms which are very dependent on their parents for survival, to more precocious species that are capable of independent locomotion at birth. Interestingly the altricial forms generally have faster growth rates. Possible evidence for slower growth rates in enantiornithines comes from fossil embryos of *Gobipteryx*. These are remarkably ossified, suggesting either slow rates of growth or maybe precocious development[1]. It is possible that the patterns of growth of enantiornithines might even have been unlike any living bird, for growth rings have been discovered in *Patagopteryx*, implying episodic growth[16].

Because growth is related to temperature, the presence of periodic growth raises the possibility that these early birds may not have been endothermic, like their modern-day descendants. However, whether or not they were ectotherms is another question. It is possible that, like some dinosaurs, they were intermediate. However, some workers have suggested that endothermy evolved before the evolution of feathers. However, as Luis Chiappe has pointed out[1], downy hatchlings are not endothermic – they cannot generate their own body heat. Chiappe believes that it is therefore possible that full endothermy evolved after the evolution of feathers and flight, and that these early birds were not only intermediate between dinosaurs and modern birds in their morphologies, but also in their metabolic rates.

So, if birds as a whole, like flowering plants and insects, took off with a vengeance during the early part of the Cretaceous, what was the effect of the great mass extinction event at the end of the Cretaceous that wiped out the non-avian dinosaurs? Some scientists, such as Alan Feduccia of the University of North Carolina, would argue that it had a tremendous impact. He is of the opinion that there is little evidence for modern bird groups before the Cretaceous/Tertiary boundary[17]. He considers that just a few straggling shore birds survived the cataclysm to give rise to all the modern bird groups in post-Cretaceous times. However, recent evidence from 'molecular clock' determinations supports the burgeoning amount of information from the fossil record that suggests a very different story.

Alan Cooper of Oxford University and David Penny of Massey University in New Zealand have analysed the genetic distances between modern groups of birds[18]. This method works on the premise that as the species evolve, their genes accumulate increasingly more mutations. Consequently the greater the time since they diverged, the greater the genetic differences between them. Using this method Cooper and Penny found that many modern bird groups appear to have diverged long before the Cretaceous/Tertiary boundary, adding weight to the arguments that the greatest explosive radiation in bird evolution was primarily an Early Cretaceous phenomenon. Their data suggest that at least 22 lineages of modern birds crossed the boundary, having originated at least 100 million years ago. Such molecular evidence would appear to support some of the new finds that show that modern ornithurine birds were present in the Early Cretaceous.

The apparent problems that some see of reconciling the fossil and molecular data may have arisen, Cooper and Penny have argued, from the historical bias of the early fossil record of birds (and also of mammals) being strongly Northern Hemisphere focussed. The earliest fossil records of many groups, such as parrots, pigeons, ratites, loons and penguins, however, are Gondwanic. For instance, Walter Boles of the Australian Museum in Sydney has described some fossil remains from Early Eocene rocks, about 55 million years old, in Queensland as containing the world's oldest songbird (passerine). The specimens are limb bones from birds of similar size to a grassfinch and thrush[19]. The previous earliest record was from Early Oligocene (about 30 million years old) deposits in the south of France. Gondwana appears to have been the major region of diversification of flowering plants. If the evolutionary radiations of these two groups were in some ways intertwined, it would not be too unreasonable to argue that much of the Early Cretaceous evolutionary radiation of bird groups may also have taken place in Gondwana.

14

Battle of the Bulge

Why did dinosaurs grow so big?

It is the winter of 1993, and Rubén Carolini is doing what he does whenever he has some spare time – searching the creeks and gullies near his home in Patagonia for dinosaur bones. He has had good success before, finding bones of huge sauropod dinosaurs that once roamed the area, some 100 million years ago, like the huge plant-eater *Andesaurus*. But today was to be different – a day that would radically change Carolini's life. For on this day he made the most important discovery of his life – the bones of what would turn out to be the largest known meat-eating dinosaur ever discovered.

Carolini lives in a small village called El Chocón, which lies 80 km from Neuquén city, in Patagonia, Argentina. The village had been built for the workers who were involved in the construction of a large dam in the early 1970s. During its heyday, when the dam construction was in full swing, more than 5000 people lived in the village. Today, after completion of the dam and transfer of the hydroelectric scheme to the private sector, only 800 people remain. Most of them are former workers of the State hydroelectric company that supplies electricity (HIDRONOR). Rubén Carolini is one of HIDRONOR's workers.

Whenever Carolini found a fossil that he thought looked particularly interesting he would immediately contact Jorge Calvo and Leonardo Salgado, the palaeontologists at the University of El Comahue in Neuquén city. When Carolini made his find in the winter of 1993 he contacted Calvo and Salgado as usual. However at this time Calvo was away in Chicago, working on his Masters Thesis on dinosaurs – with Paul Sereno. However, Salgado and another colleague, Rodolfo Coria from the Carmen Funes Museum in Plaza Huincul, were at the university. When they heard from Carolini about his discovery, they decided to visit El Chocón with another colleague, Pablo Azar, an archaeologist, to investigate the find.

Because a number of sauropod bones had been found by Carolini at this site before, Salgado and Coria thought that this new large bone would belong to yet another sauropod. When they first set eyes on the partially covered large bone, they thought their suspicions were confirmed, as it looked for all the world like a sauropod humerus. They told Carolini that they would return the next week and begin the excavation. They were rather sceptical about it being anything particularly special because they thought that it was just another sauropod, and sauropods took a lot of work to excavate. Two different sauropod species were known from El Chocón.

The next week Salgado, Coria and a number of colleagues returned to the site, 18 km southwest of El Chocón. When they first arrived they began work to excavate what they thought was a large sauropod humerus. However, after just 20 minutes of digging they were astonished to realise that this was something quite different. What they had thought was a sauropod humerus was suddenly transformed into a big, really big, theropod tibia. With further excavation a lot of theropod teeth were revealed. The scapula was found close to the tibia. After 30 days' work both femora had also been uncovered, along with a large number of vertebrae. The skull bones were found disarticulated, spread over a small area. The entire preparation back at the university took about a year.

What emerged was the largest meat-eating dinosaur ever found, about 12.5 metres long and, when living, weighing probably between 6 and 8 tonnes. Theropod expert Rodolfo Coria and Leonardo Salgado described the finds as *Giganotosaurus carolinii* (Carolini's giant, southern dinosaur), in honour of the discoverer[1]. Carolini is now the director of the recently created Museum of El Chocón (whose main star is, of course, 'Giganoto'). The local media popularised the theropod with the name 'Carolina', its unofficial name until it was formally described. Coria has since found more bones from another *Giganotosaurus* specimen, at a higher stratigraphic horizon. Jorge Calvo has discovered a partial dentary that is even larger than the original find. The skull of this second specimen, when reconstructed, would have been almost 2 metres in length!

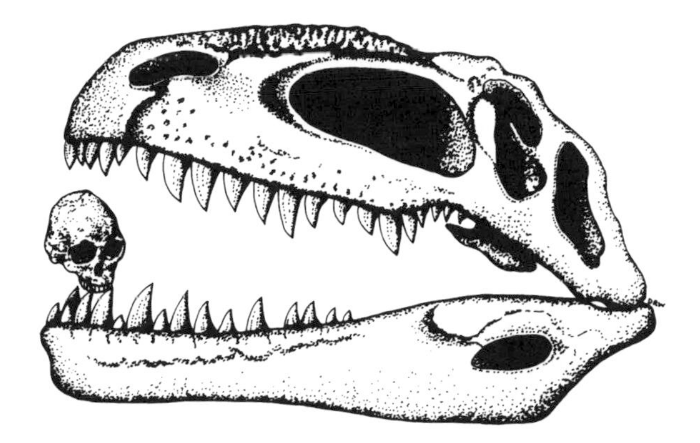

Giganotosaurus *skulls another one! The newly restored skull of* Giganotosaurus *which was found recently in* Patagonia. *It was the largest head of any predatory dinosaur known to date, measuring* 1.9 *metres long. But why did dinosaurs get so large? Changes in growth rates and times of development may have been responsible for massive growth spurts, with anatomical trade-offs sometimes being the price paid, as in the tiny arms of* T. rex.

Our fascination with dinosaurs centres almost entirely on the fact that, like *Giganotosaurus*, they were really big, scary animals. At least many of them were. Would Coria and Salgado's letter to the prestigious scientific journal *Nature* have been published, I wonder, if their dinosaur was just the size of a chicken? But not only were many theropod dinosaurs large. So too were many sauropods, hadrosaurs, stegosaurs, ankylosaurs, ceratopsians and ornithopods. But why? What underlying evolutionary mechanism was generating these Godzillas of the Mesozoic?

Until the last couple of years there had been little interest in investigating the underlying evolutionary mechanisms in dinosaur evolution[2]. In particular there has been little study of the importance of changes to the time and rate of development (heterochrony) in an evolutionary context. Surprisingly, in our view, few examples of heterochrony have been described in dinosaurs. In some ways this is understandable, because for a long time there had been a focus on just describing the huge skeletons that turned up in fossil deposits around the world. The few examples of juvenile specimens that were uncovered were often consigned to the basement to await another day. As a result, detailed knowledge of developmental patterns in most dinosaur groups was missing. Yet, current interest in dinosaur embryos has been fired by recent discoveries of eggs with preserved embryos in China and

Portugal. Furthermore, these, together with new investigations into dinosaur bone microstructure, have opened a new chapter in unravelling the evolutionary history of many dinosaurs.

Examples of paedomorphosis, the retention of 'child-like' morphologies in descendent adults, have been described in many groups of organisms. Consequently, those few souls who looked at the question of the relationship between development and evolution in dinosaurs have only looked for examples of paedomorphosis. And they were rewarded for their efforts by the recognition of some quite clear examples. Thus, for example, Bob Bakker of the University of Colorado at Boulder and colleagues have interpreted a small tyrannosaurid, *Nanotyrannus*, as having evolved by this juvenilising process[3]. However, it has recently been suggested that *Nanotyrannus* is actually little more than a juvenile *Tyrannosaurus rex*, and that the morphological differences between the two are due to developmental changes as *T. rex* grew up. Others have regarded the teeth in some hadrosaurs as being paedomorphic, while some characteristics seen in huge adult sauropods have been recognised as being characteristics found in the juveniles of earlier prosauropods[4].

Few palaeontologists, though, have addressed the basic question of how so many groups of dinosaurs attained such a large body size. As Rodolfo Coria and Leo Salgado have pointed out, large body size was attained independently in many groups of dinosaurs. Paedomorphosis was unlikely to have been operating here. In our view variations to growth rates and the duration of periods of development in dinosaurs have been a major factor in influencing their evolution. In particular this is what accounts for their massive sizes and, in some cases, unusual morphologies.

Very little is known about how fast extinct organisms, like dinosaurs, grew. However, in recent years a number of scientists have been examining the microstructure of dinosaur bones. This is beginning to generate data that can tell us a lot about growth rates and help in understanding the interrelationships between different species. What this has revealed is that growth rates in theropods were relatively rapid. David Varricchio of the Museum of the Rockies in Bozeman, Montana has shown how changes in bone microstructure in the small advanced theropod *Troodon* reveal that this dinosaur passed through three distinct developmental growth phases[5]. It had determinate growth. In other words it grew for a certain period of time, then stopped growing. Lines of arrested growth in the bone, according to Varricchio, reflect the seasonal climate in which the dinosaurs lived. It is also possible to suggest actual growth rates and time of onset of maturation on the basis of this bone microstructure. Varricchio considers that *Troodon* may have reached maturity when it was three to five years old, when it weighed about 50 kg. If similar studies can be carried out on closely related species, then changes in growth rates and the time that growth was completed might be used to intepret evolutionary mechanisms.

Similar research on bone microstructure in other dinosaurs supports the idea that the larger the dinosaur, the longer it took to become mature. This was demonstrated by Anusuya Chinsamy of the South African Museum who showed that the 20 kg theropod *Syntarsus* reached its mature body size when about seven years old. However, Chinsamy's studies suggest that the much larger prosauropod, *Massospondylus*, took up to 15 years[6].

More support for early rapid growth rates comes from a recent study by Nicholas Geist and Terry Jones of Oregon State University on the form of the perinatal pelvic girdle in a range of hatchling dinosaurs: *Maiasaura, Orodromeus, Hypacrosaurus, Oviraptor* and *Therizonosaurus*. This study revealed that they were precocial[7]. In other words they seem to have had the ability to be mobile immediately after they hatched – that is, they could probably run before they could walk. This ran counter to ideas from earlier studies of some embryological material that showed incompletely ossified ends of limb bones, suggesting that the young were altricial – that is to say they relied upon parental care at the nest while their limb bones grew enough to enable them to walk. By analogy with modern birds and crocodilians, Geist and Jones showed that these assumptions of altriciality do not hold water. They have pointed out that the growth of the femur in a perinatal *Maiasaura* is in fact very similar to the condition seen in a two week old chicken, a thoroughly precocious bird. They conclude that dinosaurs were probably precocial, similar to modern crocodiles. Another study on *Maiasaura* bone growth has revealed the presence of a bird-like pattern of growth in this dinosaur. In addition to supporting views of a close affinity between dinosaurs and birds[8], it further supports ideas that dinosaurs also had a rapid bird-like rate of growth early in development.

Our knowledge of the early development of the huge tyrannosaurid dinosaurs is very limited. Changes in proportions of bones undoubtedly occurred, though. Furthermore, the extent of development of certain skeletal elements varied between species. Thus the jaw proportions change with growth in three theropods, *Allosaurus*, *Tyrannosaurus* and *Albertosaurus*, but to differing degrees. Significant anatomical changes

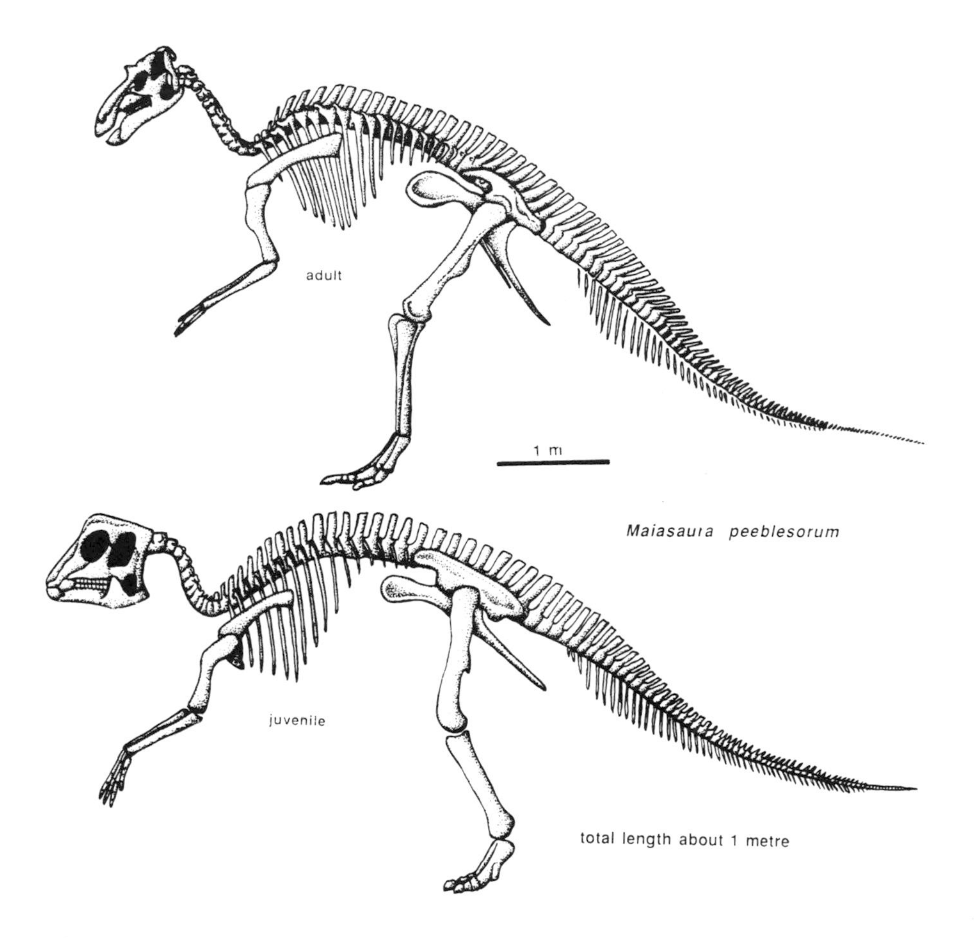

Hadrosaur family values. Juvenile and adult Maiasaura *skeletons showing major proportional changes occurring with growth. This figure first appeared in reference 2.*

also occurred with growth in the theropod astragalus, a very diagnostic bone in the foot. It is possible that there were also variations in the extent of growth between the two sexes. This has been noted in the small theropod *Syntarsus* and in the large theropod *Tyrannosaurus rex*.

Anatomical changes during development have been documented in *Tyrannosaurus bataar* from the Late Cretaceous of Mongolia on the basis of a series of skulls at different stages of growth that ranged in length from about 750 mm to 1380 mm. These changes include the transition from a long, slender skull in juveniles and small adults to a more massive, relatively shorter skull in large adults. Moreover there were changes in limb proportions, with the hindlimbs in particular becoming much more massive relative to the size of the body as a whole. In tyrannosaurids the forelimb remained very small throughout growth. Other changes during development included the closure and some fusion of cranial sutures, in particular in the lower jaw; increase in the number of serrations on the teeth; change in shape of the orbits, which became less rounded; and thickening of the metatarsals, which became more robust. The presacral column became relatively longer during development, as did the cervical neural spines. The tail, however, became relatively shorter. There was a significant increase in the size of a structure on the pelvis called the pubic boot[9].

Probably one of the more vivid images that sticks in the mind from the movie *Jurassic Park* is that of the massive head of *T. rex* screaming down on the poor soul sitting on the toilet. This demonstrates not only where not to hide when being chased by a hungry *T. rex*, but also the large head relative to body size, so characteristic of tyrannosaurids. Compared with skulls of juvenile theropods and of adults of smaller species, the larger relative size of the skull of adults of evolutionary more derived tyrannosaurids is a peramorphic feature. In other words, compared with its ancestor, this particular structure has undergone a greater degree of growth, either as a result of an acceleration in growth, or an extended growing period, or perhaps both. However, even within the skull growth rates varied, the result being a change in shape during development from a relatively long, slender head, to a more massive, deeper head. As organisms grow, not only do they increase in size, but they also change in shape. Different parts grow at different rates and for different relative durations. So, by extending the growth period major changes can occur to what the adult looks like. This is called 'peramorphosis'. Conversely, by reducing the 'amount' of growth, the adult will change shape less than its descendent and end up looking like an ancestral juvenile. This is called 'paedomorphosis'.

The same peramorphic effect of the skull growing either at a faster rate, and/or for a longer time in *Tyrannosaurus* probably also underlies the evolution of the massive hindlimbs. However, the very short forelimb and hand, in which there are only two digits, are clearly paedomorphic features. This trend is carried to extremes in the small avialian theropod *Mononykus* which has only one large digit on each hand (see Chapter 13). Such a dissociated heterochrony involving pronounced peramorphosis of body size, skull and hindlimbs and extreme paedomorphic reduction in the forelimb and hand represents a developmental trade-off arising from a metabolic balance in selection. Should any features undergo increased growth in descendent species, there is often a compensatory reduction in other areas. The tiny arms and hands of *T. rex* have been explained away in the past in many different ways. They have been interpreted as structures that helped push the dinosaur up when it fell over, to mating grappling hooks. However, in all likelihood, they served no useful adaptive purpose whatsoever. They were merely unwanted evolutionary baggage. Reductions and modifications of digits and reduced forearm size are common trends in large theropod evolution, in particular in tyrannosaurids and other derived maniraptorans, like abelisaurids. Some groups, however, went the other way and greatly elongated their forelimbs, such as in *Therizinosaurus*, *Deinocheirus* and Ornithomimidae.

The evolution of tyrannosaurids themselves from other maniraptorans shows the same sort of evolutionary trends within the group: a great growth increase in the body size, skull and hindlimbs, but a reduction in the extent of growth of the forelimb and hand. One important trend demonstrating extended peramorphic growth occurred in the tyrannosaurid pelvis. This is well shown in the development of *Albertosaurus*, where the pelvis underwent major morphological changes, especially the pubic bone. In hatchling *Albertosaurus* this bone was very weakly developed. However, in individuals 5 to 6 metres long the end of the bone swelled into a 'boot'. It became even more swollen in large adults. This pubic boot is strongly expanded in advanced adult tyrannosaurids, such as *Tyrannosaurus rex* and *T. bataar*. The oldest known tyrannosaurid, on the other hand, *Siamotyrannus* from the mid-Cretaceous of Thailand, has a rather slender pubic bone in the adult, with a pubic boot which is not as well developed as in later tyrannosaurids[10]. In its slender shape and weak pubic boot development it is like hatchlings of the later *Albertosaurus*. So just why should tyrannosaurid pubic boots get larger? One interesting possibility is that it helped the large dinosaurs breathe better. According to John Ruben and colleagues at Oregon State University, the pubic boot acts as the anchor point in crocodiles for muscles that extend from the boot to the liver[11]. When these contract air is sucked into the lungs in a piston-like manner. The larger the pubic boot, the larger the muscles, so the greater the amount of air that can be pulled in.

The indication from the analysis of tyrannosaurid bone structure suggests that their growth rate was relatively rapid. This suggests that peramorphic features in tyrannosaurids, even in large forms, were not entirely a function of delayed onset of maturation, but were also caused by acceleration in growth of specific structures. The combination of these two processes is likely to have been the key to the evolutionary diversification and success of tyrannosaurids. However, not all tyrannosaurid evolution was unidirectional. Smaller Late Cretaceous tyrannosaurids, such as *Maleevosaurus*, which is 'only' about 5 metres long, can be regarded as being essentially a paedomorphic genus. This is shown not only by its small body size, but also by its retention of the more slender snout and wider, more rounded orbit, both features of juvenile tyrannosaurids.

Similar patterns of change, in particular a dominance of peramorphic features, characterise a number of other major groups of dinosaurs. The discovery and description of embryonic and juvenile specimens of the hadrosaurid *Maiasaura*, the hypsilophodontid *Orodromeus* and the lambeosaurid *Hypacrosaurus* from the Two Medicine Formation of western Montana by John Horner of the Museum of the Rockies, David Weishampel of Johns Hopkins University and Phil Currie of the Royal Tyrrell Museum of Palaeontology in Drumheller, plus the baby dryosaurid *Dryosaurus* from the Late Jurassic Morrison Formation in Utah described by Ken Carpenter of the Denver Museum of Natural History, have enabled the nature of the early development of these groups to be established[12–14]. The bone structure of some of these young dinosaurs shows that, like the tyrannosaurids, they grew very rapidly. As they did so, there were a number of changes in relative shapes of various anatomical features.

In higher ornithopods in general a number of consistent developmental patterns can seen. The orbit decreased in size, as did the palpebral (above the eye) bones relative to the orbit. There was a relative increase in size of premaxillary and nasal bones, which eventually expanded to form the characteristic 'bill' of hadrosaurs. The snout became more prolonged in large forms and the prefrontal brow ridges enlarged, particularly in *Maiasaura* and *Prosaurolophus*. New tooth rows were added to both lower and upper jaws during development. The neural canal decreased in relative size in the vertebrae, but the neural spines increased.

The development of a long snout in adult *Dryosaurus* arose from an extended degree of growth of the nasal and frontal bones. The major cranial differences between adult hypsilophodontids, *Dryosaurus* and *Tenontosaurus*, are the result of differential growth. Thus the adults of *Tenontosaurus* possessed a long snout, smaller orbit and lacked premaxillary teeth, which are present only in hypsilophodontids within this

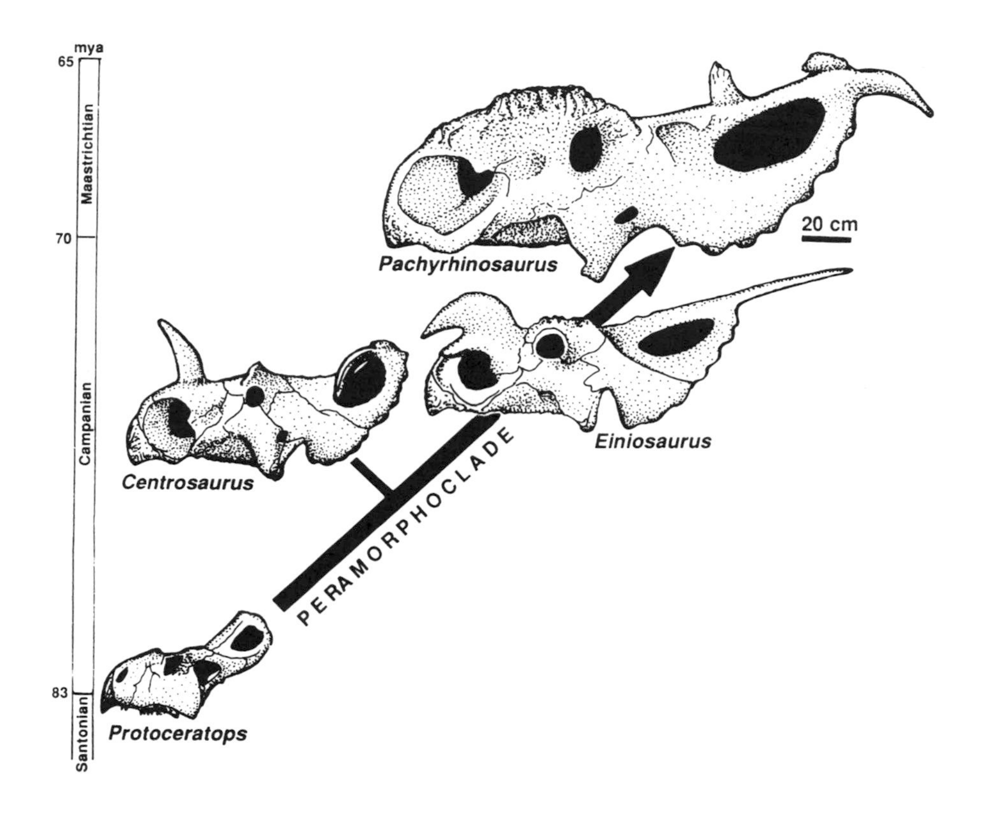

Getting big headed and horny! The major evolutionary trend in ceratopsian dinosaurs was not only to get yourself more horny (in some lineages, anyhow) but to increase your frilly ornament and enlarge your head size along with body size. This figure first appeared in reference 4.

group. They also had a long curved retroarticular process (the part at the back of the jaw that juts out) on the lower jaw, but a short palpebral bone. Amongst these cranial characteristics are a few which are apparently paedomorphic, such as the short palpebral bone. However, by far and away the most dramatic differences are the peramorphic features of *Tenontosaurus*, such as its larger body size, longer snout, development of the retroarticular process and the relatively smaller orbits.

A trend seen in both advanced lineages of stegosaurids and in ankylosaurids (thyreophorans) is the independent loss of premaxillary teeth. There are present in the most primitive stegosaurid genus *Huayangosaurus*, as well as in basal thyreophorans such as *Scutellosaurus*[15]. The same paedomorphic loss of teeth within a lineage occurs in bird evolution. Thus in *Archaeopteryx* premaxillary teeth are present, but already in some of the Early Cretaceous enantiornithine birds teeth are lost (see Chapter 13). Premaxillary teeth are only known in small ornithischians (including basal taxa such as *Lesothosaurus*), and their loss is a feature of all large members of the group that are more than 4 metres in length.

It is possible to argue from general skull growth changes that advanced ornithopod dinosaurs, like *Tenontosaurus*, are essentially peramorphic hypsilophodontids that have independently lost their premaxillary teeth and which have peramorphic growth of many facial characteristics. However the increase in growth in *Tenontosaurus* to reach a large body size (up to 7.5 metres long) corresponds with a reduction in the size of the digits and forearms. These elements are dissociated from the overall peramorphic trends in cranial characters as body size increased, indicating a similar development trade-off to that seen in tyrannosaurids.

A common feature of many living bovid mammals is the presence of horns on the skull. There is frequently a direct correlation between body size and horn size – the larger the body, the larger the horns. Either faster growth rate or prolonged preadult growth results in a larger body size and larger horns. For example, in Africa, small members of the genus *Tragelaphus*, such as the bushbuck and the sitatunga, have the smallest bodies and the shortest horns that are only slightly spiralled. The largest members, the mountain nyala and the greater kudu, sport the longest horns that develop elegant spirals. Similarly, the so-called 'Irish elk', *Megaloceros*, more appropriately known as a giant red deer, that lived during the last Ice Age in northern Europe, achieved the largest body size of its family, and by far and away the largest horns. The same pattern is seen in extinct brontotheres (see Chapter 18).

So it was with ceratopsian dinosaurs – the rhinos of the dinosaur world. Tracing the evolution of ceratopsian dinosaurs, that culminated with the massive Late Cretaceous *Triceratops*, it is possible to see a similar parallel between selection for larger body size and correspondingly larger horns. However, whether selection was primarily targeting the body size or the horn size, or a combination of both, has been, as in the horned mammals, the subject of much debate. Both in terms of size increase from little psittacosaurids through to primitive and advanced neoceratopsians, then to ceratopsians, there is an increase in morphological complexity and extent of the frill and horn arrangements. Peramorphosis has therefore been a principal factor underlying ceratopsian evolution. This is borne out by the developmental changes seen in a number of ceratopsian genera, such as *Bagaceratops*, *Leptaceratops*, *Protoceratops*, *Breviceratops* and *Chasmosaurus*. During skull development are: decrease in orbit diameter; increase in snout length (slight); slight increase in frill length, followed by subsequent shortening of the frill; widening of the frill; widening of bones around the upper jaw joint (the jugal and quadrate area); and development of the nasal horn (in *Protoceratops*).

Interestingly, the postcranial skeleton is relatively more conservative in the two advanced groups, the Centrosaurinae and the Chasmosaurinae, with the major evolutionary trends being the flattening of the digits in the large ceratopsians to cope with the greater body weight, and increase in size of a critical process on the ulna in chasmosaurines. Most morphological changes during development in postcranial elements occurred in the scapula, femur, ilium and tibia. In particular there is a thickening and broadening of certain limb elements in some later, large ceratopsids like *Triceratops*, indicating a peramorphic development to produce more massive limbs necessary to cope with the dramatic increase in body weight.

Even though the first baby dinosaur to be discovered was a specimen of *Apatosaurus* found by O.C. Marsh more than a hundred years ago, in 1883, not a great deal is known about the patterns of development in sauropodomorph dinosaurs. For instance, only a single sauropod embryo is known, a premaxilla of *Camarasaurus* from the Late Jurassic Morrison Formation in Colorado. Growth in prosauropod dinosaurs is not well documented, although very small juveniles, such as *Mussaurus* from the Late Triassic of Patagonia, are known. Comparing the skull features of *Mussaurus* with those of an adult from the same family, such as *Plateosaurus*, shows that they underwent a remarkable developmental change[4]. Their snout increased greatly in length, including an increase in anterior length of the lower jaw. With this came an increase in the number of teeth. Some of the bones of the skull, such as the quadrate, show an enlargement.

These developmental trends are continued further in the later sauropods. In particular, the skull in some groups shows a continued anterior enlargement of the jaw. Diplodocids and nemegtosaurids further continue the trend of jaw expansion in having very deep upper jaws that extend further forward of the nostrils than in other sauropod groups. The effect of this was that the nostrils were pushed upwards in some groups, such as diplodocids and brachiosaurids. Along with their massive increase in body size, these trends are suggestive of peramorphic development of the skull. Initially it showed moderate change during development in prosauropods and then extended to much greater development throughout the lineages of camarasaurids, brachiosaurids, diplodocids and nemegtosaurids.

Like prosauropods, baby sauropods are very rare. Juvenile skeletal material of *Phuwiangosaurus sirindhornae* from the Early Cretaceous of Thailand reveals that young sauropods possessed a number of features that also occur in the adults of much older sauropods[16]. This again points to peramorphic evolution of these structures. The centra and pleurocoels of the juvenile vertebrae of *Phuwiangosaurus* are very simple. The adults, on the other hand, are more complex. Older, more primitive adult sauropods from the Middle Jurassic have a relatively simple vertebral morphology, much like that in these later juveniles. Similarly, the femora of the juvenile *Phuwiangosaurus* are relatively more elongate and slender than in the adult. This mirrors the condition seen in more primitive sauropod adults. The massive increase in size of later sauropods, and the greater elongation of the necks and tails from their prosauropod ancestors clearly points again to either an increased growth rate and/or a longer period of growth. This is how these largest land animals that ever lived achieved their phenomenal size.

The ultimate question, however, must be, what drove the massive body size increase in so many lineages. What was being selected for? Was it the various anatomical structures? Or was it the overall larger body size? Or maybe it was a combination of the two? Although it has been codified as Cope's Rule, the pattern of increasing body size through time that occurs frequently in every group of animals is by no means an inviolate rule. Dave Jablonski of the University of Chicago has recently shown in an analysis of Cretaceous bivalves and gastropods that there is no clear-cut evidence for a predominance of body size change through time. Some lineages get bigger, others get smaller. What is more unusual is a lack of size change. Size changes may have been commonplace in evolution, but they were equally as often to small size as to large. In the case of dinosaurs there were certainly many small forms, and in most lineages there would be examples of smaller taxa evolving from larger forms. But the evidence to date seems to indicate a predominance of selection for larger body size.

Either way, the question still remains. Why are there evolutionary changes in body size at all, whether it be to a larger or smaller body size? The answer, as I discuss at more length in Chapter 16, may well lie

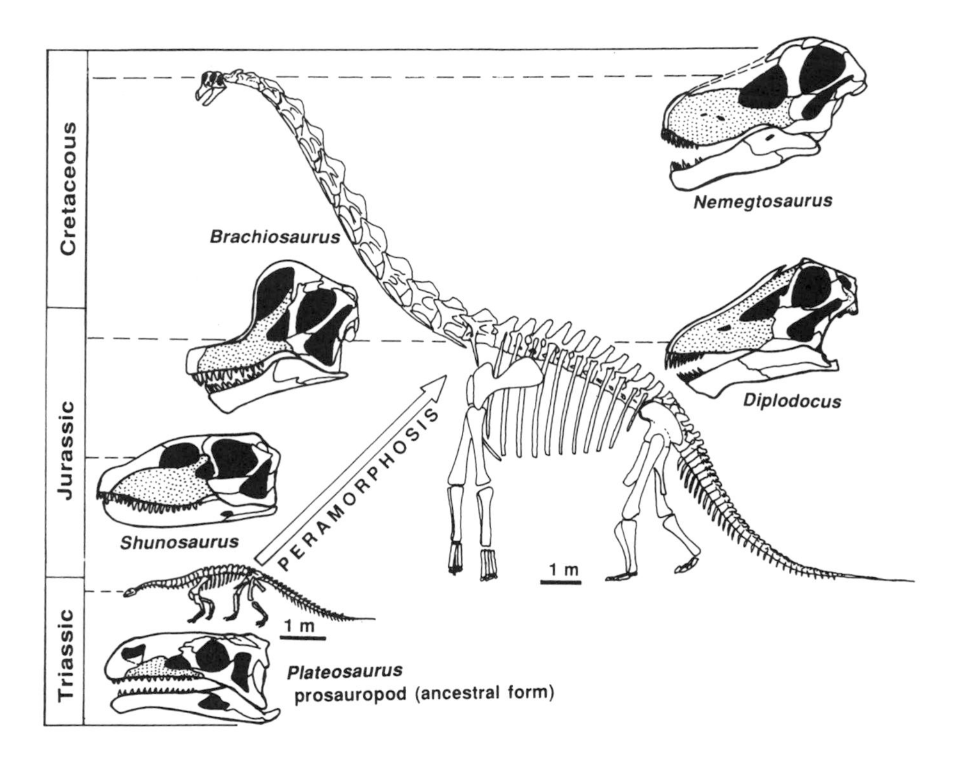

How to get really, really big. The sauropods were the largest creatures ever to walk the Earth, reaching estimated lengths of nearly 40 metres and weights of up to around 100 tonnes. Their rapid rise to be king heavyweights of the animal world came about through lengthening growth periods and accelerating rates of development. This figure first appeared in reference 4.

with predation pressure. For to exist at all we must eat and avoid being eaten. Modern ecological studies have demonstrated how many predators operate a so-called 'optimum foraging strategy'. What this means is that they will seek out prey that is within a certain body size range. Too small, and it is not worth the effort of chasing it, in terms of the rewards in energy that would accrue. Too large, and the energy expended may be greater than would be accrued. The result of this is that selection will favour target prey species that lie outside of this optimum foraging range. Smaller species can also hide more easily. This way selection will always be targeting those body size ranges at the extreme of the variation within the species. This is obviously complicated if a species is preyed upon by more than one predator, and these two have differing optimum foraging strategies. In the case of dinosaurs it could be argued that the selection has targeted those herbivorous species that could grow faster, or for longer, and attain a larger body size, making them less susceptible to predation pressure. Obviously for the juveniles, thc faster they can grow and attain larger body size, the greater their chances of survival.

An evolutionary arms race probably existed between dinosaur predators and prey. For as the prey species became larger, so too there would be selection pressure favouring those larger predators that can tackle the larger prey species. This evolutionary ratcheting of the (I hardly dare to use the phrase) ‘companion’ species, is therefore likely to have led to predominance of trends to increased body size. The limit would have been reached when physiologically the organism became unviable. Clearly these strategies worked well for many groups of dinosaurs, as they existed for more than 150 million years before succumbing to extinction at the close of the Cretaceous. Maybe in the battle of the bulge they just got too big for their boots. Who knows? Now, what caused that is another story . . .

15

The Tooth, the Whole Tooth, and Nothing but the Tooth

Gondwana's treasure-trove of fossil mammals

If you looked up the scientific diagnosis of the African elephant you would imagine that it might mention something about big tusks, a long trunk and even the great lumbering mass of the creature. A not unreasonable expectation. But nothing could be further from the truth. All that concerns the pachyderm taxonomist are the intricate ridges and grooves that grace an elephant's tooth! Much of our understanding of ancient mammals is dependent on the most resistant part of the animal – its teeth. For after a mammal dies, the soft parts are the first to decompose, followed by the bones. Last to go, if they go at all, are the teeth – incredibly resilient structures.

Mammalian teeth are easily characterised by their complex crown morphology and often multiple root structure, and the unique structure of the enamel in having very dense, closely packed, and well-ordered crystallites. Although much of the history of mammalian evolution is based on well-preserved fossil skeletons, many complete with skulls, from most of the world's continents, the one continent whose mammalian evolutionary history is still shrouded in mystery is Australia. This is despite the fact that the continent still carries on its back today the relicts of some of the earliest mammals, in the form of two living monotremes, or egg-laying mammals. The story of the hunt for Mesozoic mammals in Australia is one of the great searches in palaeontological history. It is a story that began back in the 1950s and has continued for four decades, costing thousands of research hours. This search, for early, Mesozoic mammals, was ultimately fulfilled in 1985.

For more than a century the known fossil history of mammals in Australia was restricted to the last 25 million years of geological history – truly well past the time of the start of mammal's evolutionary race, when all the players were on the track and coming into the final straight[1]. The real story, that is of mammal origins and distributions within Gondwana, remained a mystery, until a few mysterious molars and jewelled jaws of opal were discovered in the 1980s and another in 1997. These have given us tantalising insights into the strange world of Gondwana mammals that scampered beneath the feet of dinosaurs.

Before embarking on this Gondwanan story, let us first gaze at the global picture of mammal evolution to see why Australia's 'mammalian missing links' are so important. Mammals first evolved back in the Triassic Period, at roughly the same time as the first dinosaurs appeared, about 225 million years ago. The oldest mammals appear not to have been too different from the advanced mammal-like reptiles that abounded in the Late Permian and Early Triassic. The only substantial way in which they differed was in the lower jaw joint. This had shifted slightly giving the characteristic mammalian arrangement whereby the lower jaw articulates with the squamosal bone (one of the cheek bones) of the skull. This, together with further differentiation of the dentition into different types of teeth – molars, premolars, incisors and canines – is what made an early mammal differ from an advanced reptile. Like the elephant, it was all in the teeth.

The Jurassic and Cretaceous Periods saw the rapid rise of dinosaurs, dominating the terrestrial habitats of the Earth, whilst mammals flourished in their shadows. It is a fallacy to think that mammals were necessarily smarter than dinosaurs, or that mammals eventually *overcame* the dinosaurs and played a major part in bringing about their extinction. As far as we can tell from the fossil record, mammals were simply in the right place at the right time, and just lucky to get a break when the dinosaurs died out. Many niches were made vacant, so the few mammal families around at the time simply moved in, along with a few strange (and flightless) birds, and radiated.

It is Australia, back in the early 1980s. The search for Mesozoic mammals is in an intense phase as Tom Rich and Pat Vickers-Rich, of the Museum of Victoria and Monash University respectively, continue their regular summer digs in the Early Cretaceous sandstones of coastal

Victoria. This is after many other attempts to find Mesozoic mammals in Australia have failed. These include Rich's expeditions to the extensive Cretaceous outcrops of northern Queensland and Western Australia, their trips and those of Alex Ritchie of the Australian Museum to the opal fields of Lightning Ridge and Coober Pedy, and even earlier attempts by Mike Archer of the University of New South Wales and Ernie Lundelius of the University of Texas at Austin, at sieving the Late Cretaceous greensands and clays of Western Australia. As we shall see, persistence and downright pig-headedness were to eventually pay dividends.

But long before they made their lucky strike, one of Australia's major palaeontological discoveries was made, as usual in a somewhat serendipitous fashion, in 1984 by Alex Ritchie. He was in Lightning Ridge to inspect a collection of opalised fossils that the Australian Museum was interested in purchasing from the Galman brothers, local opal collectors and dealers. In a typical outback motel room in the heart of Australia, the search for Australia's first Mesozoic mammal came to a rather unspectacular climax. The odd collection of opalised fossils was laid out on a bed by the Galmans. Alex Ritchie's eye was suddenly caught by one small specimen that was sitting in a tiny box. It was a jaw with three complex teeth in it. The hairs on the back of Ritchie's neck promptly stood to attention, for he realised that nestling in this box was Australia's first Mesozoic mammal. Just a small jaw containing three molar teeth, perhaps, but it was more than 85 million years older than the country's previous oldest known mammal.

Fortunately the Australian Museum purchased the collection for $80 000, money raised by sponsorship and donations. This wonderful, opalised jaw on its own justified the expense. Soon after it was named *Steropodon galmani*, meaning loosely 'Galman's lightning tooth'[2]. *Steropodon* was diagnosed as being a monotreme, one of the egg-laying mammals, and a close ally to the duck-billed platypus. Ironically, platypus have no teeth, so how was the link made to the fossil jaw? The answer to this lies in the embryonic development of the platypus, when the baby develops its milk teeth. These juvenile teeth have a close resemblance to those of *Steropodon*. Initially *Steropodon* was placed, questioningly, in the same family as the platypus, the Ornithorhynchidae. However, later, after another Mesozoic mammal was found from the same region, it was elevated to its own family, the Steropodontidae[3].

Since this remarkable find, Alex Ritchie has regularly returned to Lightning Ridge to attend local opal shows, setting up a stall where he

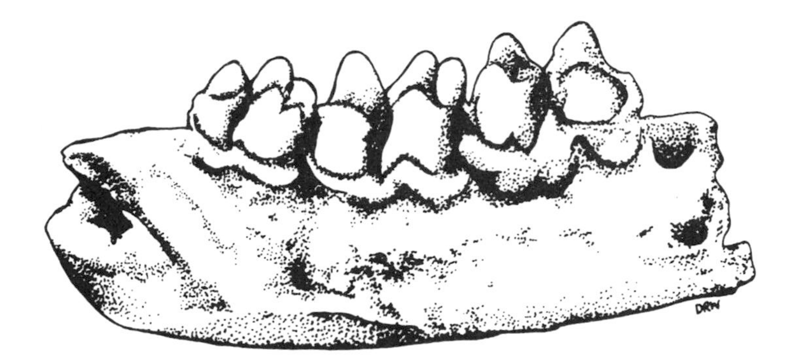

What lovely teeth you've got, grandma! This lower jaw of Steropodon, *preserved in opal, was the first fossil* Mesozoic *mammal to be recognised from* Australia. *It caused a scientific sensation when first identified as belonging to the egg-laying monotremes, possibly a close relative of the platypus. Incidentally it was also one of the largest known mammals of its day, being cat-sized.*

encourages opal miners to bring him their fossil treasures from the underground labyrinth of the opal diggings. Success struck again about 10 years after the first find, with the acquisition of three more Mesozoic mammal jaws. Two of these lack teeth, but the other is a superb specimen with teeth distinctly different from those of *Steropodon*. These teeth are rounded and look very much like hot-cross buns. Mike Archer, who was to study it, was initially tempted to name the new beast 'Hotcrossbunodon', but sanity prevailed and so instead it was named *Kollikodon ritchiei* (from the Greek meaning 'hot-cross bun tooth'!) Alex Ritchie's important role in finding these jaws was rewarded with the species being named after him[3].

From these two small jaws our ideas on early mammal evolution have been dramatically changed. They demonstrate the radiation of monotremes that must have occurred in Australia during the Mesozoic. An even greater upheaval in Gondwana mammalian evolutionary history is likely to occur with the recent discovery in 50 million-year-old sediments in Murgon, Queensland of a possible eutherian (non-marsupial) tooth, which we discuss below. Even more amazing is the recent discovery of a fossil platypus tooth in South American sediments.

Monotremes are the most primitive of all living mammals, not only because they retain the primitive reptilian method of laying eggs rather than giving birth to live young, like placental and marsupial mammals. But their skeletons also show many relict features of more primitive animals. If you look closely at the only two living forms of monotremes, the duck-billed platypus (*Ornithorhynchus*) and the long-snouted spiny echidnas (*Tachyglossus* and *Zaglossus*), you couldn't find two animals more dissimilar in appearance and lifestyle. The platypus is an aquatic animal, with duck-like bill covered in a unique sensory system, webbed clawed feet (males with venomous spines), and flat beaver-like tail. It inhabits the streams of eastern Australia. The two echidnas, one from the forests of New Guinea (*Zaglossus*), the other from Australia (*Tachyglossus*) are land dwellers that feed exclusively on ants and termites. The discovery of *Steropodon* and *Kollikodon* makes it clear that monotremes once ruled the mammalian roost in ancient Australia[1].

More conventional types of platypus occur as far back in the fossil record as the Early Paleocene, some 60 million years ago. But this fossil was not found in Australia. It came from Patagonia, at the bottom end of South America. Its discovery caused quite a stir in the palaeontological world, because its announcement in the scientific journal *Nature*[4] was hot on the heels of an earlier, equally unexpected announcement made by Henk Godthelp and colleagues at the University of New South Wales[5]. The bombshell that they dropped was that a South American-type placental mammal (i.e., neither a marsupial nor a monotreme) had been found in sediments of similar age near Murgon in Queensland, Australia – God[t]help us all! Our cosy concept of marsupials surviving and thriving in Australia, because of the failure of eutherians to colonise Australia, had seemingly been shattered, along with long-held views on the early distribution patterns of primitive mammals, based essentially on finds from North America, Asia and Europe.

Sixty million years ago Gondwana was in its final death throes, as Australia and Antarctica were straining at the leash, desperate to rift apart. South America had already broken its apron strings and drifted off. It would seem that a unique kind of mammal fauna had already been established prior to these events, right across the southern and eastern provinces of Gondwana (Antarctica, Australia and southern South America).We can imagine these cold, but lush forests of southern Gondwana as home to a diverse range of furry creatures, including the platypuses, primitive marsupials, like opossums and small predatory forms, the first bats, and a range of eutherian or placental mammals, including the enigmatic fossil from Murgon named *Tingamarra*, which was possibly a member of a group of mammals known as condylarths. These were a group of primitive, plant-eating mammals about the size of a fox-terrier, that evolved at the end of Cretaceous, just before the demise of the dinosaurs. They are believed to have been the ancestors of many of the more recent plant-eating mammals[5].

Condylarths were a group of herbivorous eutherian (placental) mammals that dominated the basal Paleocene mammal fauna in South America. It is not easy to specifically characterise these animals. They may represent an evolutionary 'grade of organisation' rather than being a discrete group, although several lineages of condylarths are readily definable. From these condylarths emerged the major mammalian group called ungulates, or hoofed mammals. These are the artiodactyls (pigs, camels, deer, antelopes, cattle), the perissodactyls (horses, rhinoceroses, tapirs), various South American ungulates, possibly elephants and maybe even whales, porpoises and dolphins. Named *Tingamarra*, after the site of its discovery, the Murgon fossil is based on a single tooth which its discoverers considered could be a condylarth placental. This caused a huge stir as it seemed to be the first terrestrial non-marsupial or monotreme mammal to occur in Australia, other than those brought on to the continent by humans in much more recent times.

Then in early 1997 the Richs' persistence paid off with one of Australia's most exciting fossil mammal discoveries. A jaw was found near Inverloch in eastern Victoria in 115 million-year-old Cretaceous rocks. This once belonged to a mammal having no obvious similarity to any other previously known Australian mammal. Tom Rich and colleagues named this beast *Ausktribosphenos nyktos*, meaning the 'Australian tribosphenic Cretaceous mammal that lived at night'[6]. In short this means that the tooth exhibits features thought to be primitive for both marsupials and eutherian mammals. The 'tribosphenic' tooth has a triangular-shaped basin which forms the effective biting or occlusal surfaces adapted for both cutting and crushing food at the same time.

The discovery sent shock waves through the world of fossil mammal research for the simple reason that Rich and his co-workers identified this jaw as provisionally belonging to a placental (eutherian) mammal. As such it would make it one of the oldest representatives of that group and certainly the first occurrence of these mammals in the Mesozoic of Australia. Its occurrence in Australia also upsets the apple cart because placentals are simply not represented here in the fossil record (apart from the controversial tooth of *Tingamarra*, and bats). However, almost before the story was dry on the pages of the journal *Science*, Mike Archer, of the University of New South Wales, was delivering a keynote address at a palaeontological meeting in Woolongong in December 1997.

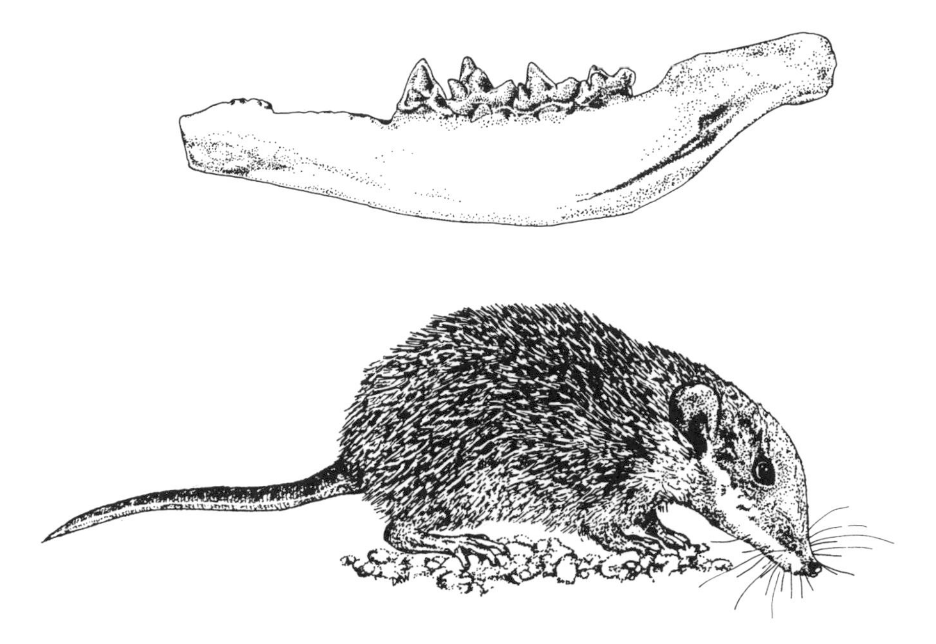

Another ugly fact to ruin a beautiful theory? Conventional wisdom says placental mammals, like rats and mice, didn't reach Australia until a couple of million years ago. This jaw, of Ausktribosphenos *might be just such an ugly fact. Its discoverers believe it may have been a placental. The only trouble is it's about* 115 *million years old!* Reconstruction after a drawing by D. Gelt.

Archer, discussing the origins of mammals in Australia, offered a very different interpretation of the little jaw of *Ausktribosphenos*. He argued that the jaw was not from a placental at all, because it had teeth that he considered were more likely to have given rise to the monotremes. In his opinion *Ausktribosphenos* is thus a primitive sister taxon to the monotremes and not related to the basal radiation of placentals. The debate rages on . . .

The discrete features of each mammalian tooth hold the key to its ancestry. The three major surviving mammalian lineages (monotremes, marsupials, eutherians) have tooth morphologies that are distinctive to the experts who study them. However, the interpretation of patterns of mammalian evolution, based solely on teeth, has caused fierce debate. *Tingamarra*, for example, was identified and named from one lower right molar about 2 mm long. The ultrastructure of its outer enamel layer is made of horseshoe-shaped prisms roughly 3 micrometres in diameter, arranged in a pattern that is said to be unique to eutherian mammals. If we look at the hills and valleys of its tooth shape, then it is missing the characteristic features always claimed to be found in marsupials, known as a well-developed buccal postcingulid and antero-posteriorally compressed trigonid. These are the specific cones and linking ridges between them on the tooth. Their absence is fuel to arguments that this tooth is not from a eutherian mammal at all. It has been argued[7] that the morphology of this single tooth could be interpreted either way – as a primitive marsupial or as a eutherian. So the jury is still out on whether or not condylarths once roamed Australia during the Tertiary. Whether it will survive the rigours of palaeontological analysis and survive as a eutherian is also another question.

The little platypus tooth from South America is an upper right second molar measuring about 8 mm. This is very large for a monotreme, being slightly bigger than the Cretaceous form *Steropodon*. When compared side by side with the same molar tooth from a 30 million-year-old Australian platypus, *Obdurodon*, the tooth is amazingly almost identical! In this case there is no controversy. All agree that this tooth once sat in the mouth of a monotreme. Yet it poses some major questions on the radiation of monotremes. Apart from these scant amounts of fossil data, evidence from DNA hybridisation experiments suggests that the two main stocks, the echidnas and the platypuses, diverged some time between the Late Cretaceous and the Early Tertiary, between the times of the Australian Cretaceous *Steropodon/Kollikodon* and the South American Tertiary *Obdurodon*[4]. However, before investigating the mechanisms whereby monotremes and marsupials first came into Australia, we must first look outside of our continent at the early mammal faunas of Antarctica and South America.

Fossil mammals from Seymour Island, which is an island in the Weddell Sea, lying off the coast of and near the northern tip of Graham Land (Antarctic Peninsula), dated at around 40 million years old, indicate that marsupials cohabited with a variety of placental mammals. The fauna includes microbiotheriid, polydolopid and derorhynchid marsupials, plus edentates and ungulates. These groups are typical of the South American marsupial fauna from slightly older times, suggesting that the Seymour Island marsupials were separated from the mainland South American marsupial fauna after the early part of the Eocene Period, about 40 million years ago. The real question is when mammals invaded Australia, and how long an open freeway may have existed between South America, Antarctica and Australia.

While Mesozoic marsupial remains have been found in North America, to date none has been found in Australia or South America. It is possible that marsupials may have dispersed into Australia prior to the end of the Mesozoic Era, and that the land bridges connecting Australia to South America did not exist during the early part of the Tertiary. But was the dispersion one of what has been called 'open dispersion' or 'filter route dispersion'? An open type of dispersion implies a free interchange whereby animals from both sides could freely migrate back and forth. A filter route dispersal mechanism, however, involves a more restricted type of distribution whereby animals might disperse part of the way then be restricted by severe climatic or geographic barriers, thus not allowing complete interchange between the two regions.

In the case of the Australia–Antarctica connection at the close of the Mesozoic Era and the Early Tertiary, it appears that there may have existed a filter route over a very narrow dry land connection, at high latitudes. There may have been a one-way migration of mammals from Antarctica, but no reverse passage for any native animals then existing in Australia, such as the monotremes. However, the recent discovery of *Obdurodon*, the platypus from the Paleocene of South America, suggests that monotremes may have been widespread across Gondwana. However, only in Australia have they persisted to the present day, following the breakup of the southern supercontinent. While the absence of marsupials in Australia and South America in the Mesozoic argues for a Northern Hemisphere origin, the discovery of a single tooth, in the same way that the Cretaceous monotreme's jaw turned up, would blow such a theory sky-high. The question still remains as to why, with the possible exception of *Tingamarra*, eutherians didn't populate Australia. Clearly, to solve this conundrum what we need is the tooth, the whole tooth and nothing but the tooth!

The same problems that we have encountered with dispersal of vertebrates across the rifting Gondwanan southern land bridge, were also mirrored in the Northern Hemisphere diversification of mammals and birds. The classical picture is one of birds and mammals suddenly arising out of the ashes of the dinosaurs and waking up to the wealth of new niches suddenly available, then going for it like the proverbial bat out of hell. The fossil record indicates that the beginning of the Tertiary Period saw the very rapid diversification of mammals and birds, perhaps in as little as one million years, into all the major niches they occupy today.

If we put the evidence from fossils aside, however, and look at the problem from a different perspective, some very interesting results can be gleaned. Recently estimates of molecular divergence times for several of the major orders of living placental mammals and some birds have been made[8]. This technique is based on the assumption that genes evolve at a reasonably reliable rate so that the degree of similarity in the amino acid sequences of proteins or the DNA that codes for them can be used to estimate when major lineages evolved. The more they differ, the longer ago the groups diverged. Their results point to a series of divergence events between about 110 and 90 million years ago This is despite clear absences of some groups from the fossil record. This is particularly so for three separate lineages of therian mammals.

The Primate–Artiodactyl divergence is estimated to have occurred about 100–90 million years ago; the Primate–Rodentia divergence about 90 million years ago, and the Rodentia–Artiodactyl divergence at 110–100 million years ago. The implications for all this are simple – that the divergence times for the separation of these lineages are about 50–90% earlier than the fossil record suggests. If so, then the break up of Gondwana at the end of the Cretaceous may well have played an important role in determining the present-day distributions of many of these orders, especially mammals. The fossil record shows that most modern orders of mammals appeared near the beginning of the Tertiary Period. Yet just where they evolved from has often been a mystery. Novel approaches using molecular techniques give the palaeontologist a working hypothesis to test – a target to go out and shoot down, or give some support to as new discoveries are made.

Recent work on the group of Mesozoic mammals known as multituberculates puts a new perspective on mammalian evolution that is not centred on teeth alone. The name 'multituberculate' derives from the complex nature of the molar teeth which bear many rows of uniform cusps. The group had a timespan ranging longer than most of the better known mammal groups, that is from the Late Jurassic through to the Eocene, even surviving the end Cretaceous extinction event. The new discovery of an articulated multituberculate mammal, *Bulganbaatar nemegtbaataroides*, from Mongolia by Paul Sereno of the University of Chicago and Malcolm McKenna of the American Museum of Natural History, has shown that although these animals retained a primitive bone, the interclavicle, in the shoulder girdle, in other ways the shoulder girdle is very advanced for a Mesozoic mammal[9]. The interclavicle is a median bone in the shoulder girdle that goes back as far as the lobe-finned fishes and is retained throughout tetrapod evolution, being lost in higher mammals. In the past multituberculates were grouped with the primitive egg-laying monotremes and some other extinct Mesozoic mammal groups, so the evidence presented by Sereno and McKenna shows that the advanced mobility of the multituberculate shoulder girdle is a feature they share with the therian mammals, the group to which most modern groups, such as primates, rodents and artiodactyls, belong.

At least six characteristics of the postcranial skeleton in the Mongolian multituberculate are shared with modern therian mammals. Thus, as Sereno and McKenna point out, the evolutionary relationships of the early mammalian groups based mostly on teeth or some cranial characteristics cannot be adequately resolved. This serves as another example of where the tooth, the whole tooth and nothing but the tooth can result in conflicting hypotheses which ultimately require better, more complete material to resolve the conflict. Luckily, in the case of multituberculates, the superb new material from Mongolia was capable of doing just this.

One of the really major evolutionary transitions that took place during mammalian evolution occurred during the Early Tertiary. This too is another example of where teeth were not of central importance to macroevolutionary trends. Much has been written on the famous story of the horse – evolving from five toes down to one as it grew bigger and faster (and we contribute even more to this story in Chapter 18). But what of really major morphological transformations? The biggest and most stunning is without doubt the evolution of whales. Unlike the fish–tetrapod transition, we now look at the adapation from land to water, involving the evolution of new feeding mechanisms, new sensory mechanisms, major changes in size, and other major morphological changes to the standard mammalian body plan.

New discoveries of fossil whales from the Eocene of Pakistan made by Phil Gingerich of the University of Michigan and his colleagues, has filled in some of the vital gaps in the story of their ancestral group, the mesonychids – how they left the land and became fully aquatic whales. The earliest known whales are the archaeocetes, meaning 'ancient whales'. Like their ancestors, the hoofed mesonychids, they had long skulls, with large triangular molar teeth. *Pakicetus*, the oldest known whale, was a small archaeocete with a much smaller skull and auditory bulla (for hearing) than any of the other four known Early–Middle Eocene primitive whales. *Pakicetus* lacked also the specialisations for effective hearing in water found in the other whales, and as such it is generally regarded as the most primitive known whale. However, it remains poorly understood[10].

The discovery in the 1990s of *Ambulocetus*[11], and more recently *Rodhocetus* from the Early to Middle Eocene of Pakistan[12], along with an earlier form, *Pakicetus*, of basal Eocene age, shows the transition from a running, terrestrial mammal through to fully aquatic cetaceans that swim with spinal undulation of the tail flukes. *Ambulocetus* had well-developed arms and legs and a long tail. It was probably capable of limited terrestrial locomotion, but in water was thought to have swum by flexing its vertebral column up and down, in the same way as modern otters[13]. This is shown by the shape of the lumbar vertebrae. The feet were large and could have assisted in swimming. The back muscles primarily powered the rear limbs, as they do in phocid seals today. On land the animal was much like a seal, having sprawling hands and feet. *Ambulocetus*, was clearly a whale, though with several specialisations of the skull found only in early archaeocetes.

Rodhocetus, described by Gingerich and colleagues in 1994, was the first discovery of an early whale with the complete lumbar, thoracic and sacral regions of the vertebral column preserved. As such it gives a special insight into the nature of how cetacean swimming began. The high neural processes on the anterior thoracic vertebrae, along with the pelvis which articulated directly with the sacrum, are features otherwise only encountered in terrestrial mammals that support their body weight with their limbs. This suggests that this early whale retained some ability to move about on land. The neck vertebrae are short and the skull is very long as in later whales. The unfused sacral vertebrae would have enabled the creature to flex its spine up and down more easily, again, much as in modern whales. The remains of *Rodhocetus*, one of the more complete early whale fossils, came from deep-shelf sediments, suggesting it was the earliest whale to venture well out into the open sea, and not be restricted to shallow, nearshore habitats. Also, like modern whales, *Rodhocetus* had the functional adaptations required for major flexing of the body using the back and abdominal muscles. This is shown by the presence of unfused sacral vertebrae, high neural spines on the dorsal vertebrae, short, stout anterior caudal vertebrae and long ventral chevrons. Thus Gingerich's team has found an almost perfect intermediate mammal between the mesonychids and the true whales.

As whales continued to evolve through the Eocene Period, they increased rapidly in size, reduced the hindlimbs to just remnants, lost their tail, and their front limbs developed more into powerful flippers for directional guidance and manoeuvrability. Other whales of this time were experimenting with various locomotory systems. *Indocetus*, for

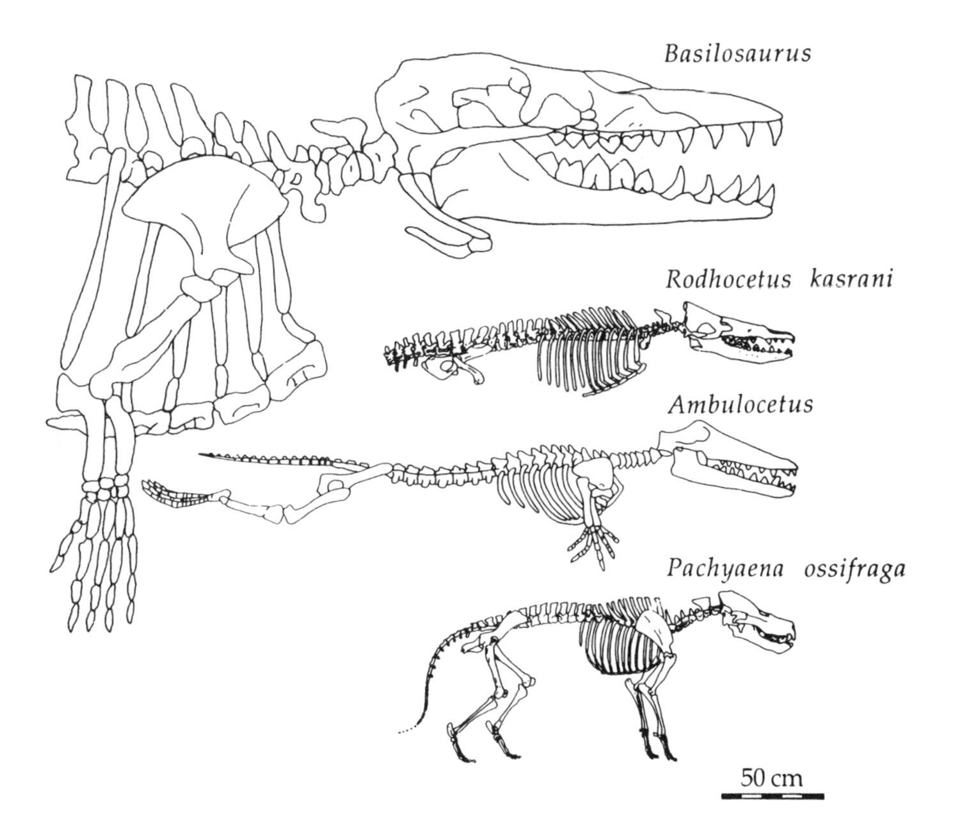

Thar she blows! Moby Dick has its earliest ancestors dating back some 53 million years ago to the dawn of the Eocene Period. Mesonychid mammals like Pachyaena *which frequented the seas to hunt probably gave rise to the first aquatic swimming forms, whales like* Ambulocetus *and* Rodhocetus. *However, the first true behemoths of the sea appeared not long after them, as seen by* Basilosaurus, *reaching nearly 18 metres in length.*

example, had long hindlimbs and a fused sacral region, comparable with fully terrestrial mammals, but it inhabited shallow marine environments. *Protocetus*, on the other hand, had a sacrum that did not articulate with the pelvis, but is still thought to have been fully aquatic, incapable of supporting its weight on land. These remarkable series of fossils show us that the mesonychid ancestors of whales took gradually to the water, and through an amphibious seal-like locomotory phase, eventually graduated into wholly aquatic animals. Following their successful occupation of this new niche, a whole array of different whale types evolved. Amongst these were the first giants of the animal world, the mighty baleen whales.

Archaeocetes, like the primitive Eocene whales, were primarily predators armed with large serrated teeth and long, stout jaws. The most specialised lineages of whales were the two modern lineages, one leading to baleen whales (filter feeders), the other to the odontocetes, or toothed whales, which include sperm whales, killer whales, dolphins and porpoises. The archaeocetes came to an end at about the same time as these two modern lineages diversified, at the end of the Oligocene Period. During the Miocene representatives of many of the modern genera, or at least their families, had appeared, including *Baleanoptera*, the great blue whale.

The origin of baleen whales from toothed archaeocete whales has been elucidated in recent years from new finds and recent research by Ewan Fordyce of Otago University in New Zealand. Ewan and I were both at Monash University together in the early 1980s. He was a postdoctoral fellow working on whales, while I was a postgraduate student working on fish. At this time Ewan often searched for fossil whales in the coastal limestones of western Victoria. He collected some superb material from near Torquay of an enigmatic form called *Mammalodon colliveri*. This animal was then thought to be an archaeocete whale, as it had typical archaeocete-type teeth. On the basis of a skull and lower jaws, plus several vertebrae, Ewan came up with a new picture of this strange whale as being a sort of missing link between archeocetes and baleen whales. His evidence for this was that although *Mammalodon* has teeth, they were widely spaced apart and had high gum lines, suggesting there was a large amount of lip or skin tissue between the tooth gaps. Furthermore, the lower jaws in toothed whales meet at strong symphyses to give strength to the bite, yet in filter-feeding baleen whales the lower jaws are loosely connected to allow some rotational movement of the long jaws when catching krill or plankton. *Mammalodon* possessed the latter condition, of a weak median symphysis. The skull showed that it had a more baleen whale style of echolocation, and thus it was deemed to be more of a primitive baleen whale (still retaining teeth) than a typical, specialised archaeocete. The widely spaced teeth of *Mammalodon* may have functioned like those of a crab-eater seal, which sieves food out of the sand using its serrated triangular teeth. Eventually the use of soft tissue between the teeth would have become more efficient for filtering smaller particles, and baleen would eventually have replaced teeth entirely.

The enlargement of body size seems to accompany all filter feeders throughout time, such as the giant whales and basking sharks, the huge titanichthyid and homosteid placoderms (Devonian age), and even the huge 16 metre long Jurassic teleostean fish *Leedsichthys*. Despite the new fossil discoveries having revealed the transitions from land mammals to whales and even from toothed whales to baleen whales, the evolutionary processes involved in these major morphological transformations have not been adequately discussed in the literature. Clearly, the rapid increase in size of whales, like dinosaurs, is due to accelerated or longer periods of growth, yet at the same time, there was a reduction in the size of certain organs, in this case the hindlimb and pelvic skeleton. Other skeletal elements, however, underwent a massive prolongation, such as the enormous lower jaws in baleen whales and the enlarged frontal bones. Like some dinosaurs, such as *Tyrannosaurus rex* (see Chapter 14), evolution of massive body size was accompanied by some evolutionary give-and-take: developmental trade-offs that saw some anatomical features increase greatly in size and complexity, while others were greatly reduced or failed to develop at all. What you gain on the evolutionary swings, you lose on the evolutionary roundabouts.

16

Murder and Mayhem in the Miocene

Predation pressure driving evolution

The main aim in life for most animals, on a day-to-day basis, is to stay alive. This they do in two basic ways – by eating and by avoiding being eaten. The stupendous variety of shapes and lifestyles that animals have evolved over the last half a billion years or so, bears testimony to this. As I pointed out in Chapter 3, even in the Early Cambrian top-line predators, like *Anomalocaris*, were much in evidence, spending their lives cruising the seaways looking for tasty morsels. The evolution of many life history strategies has been focused on minimising predation pressure. Thus the rarity of infauna (animals that burrowed into the sediment) in the Early Cambrian soon changed as more and more groups found ways of hiding in the sediment, as a means of escaping from becoming something else's dinner.

Much of the evolution of fishes was influenced by the need to eat, or avoid being eaten. The colonisation of land itself may likewise have been predator driven; those animals with the ability to survive on land, away from the fangs of hungry predators, were more likely to endure (given the possession of the right survival equipment for a life on land). So too the conquest of the air by insects, reptiles, birds and mammals. If you are small and likely to be eaten, but you can get up and away to soar high above your attacker, you may well survive just long enough to pass your genes on to the next generation.

In the sea, many evolutionary trends, which have led to the colonisation of new marine environments, may well have been propelled by predation pressure. The evidence for this comes from a fossil record that is dominated by marine invertebrates. In particular, I want to look at the impact of predation pressure on the evolution of sea-urchins.

After a diet in preceding chapters of huge marine reptiles, dinosaurs and spectacular mammal fossils, a bunch of fossil sea-urchins might seem a bit tame. However, they have provided an excellent investigative tool to elucidate the nature of evolutionary mechanisms, evolutionary trends and the driving force behind evolution. While it is not too difficult to observe the ravenous hordes of animals that attack and eat urchins today, we must turn to the fossil record to see how sea-urchins have evolved different strategies to avoid being eaten. For unless they keep one step ahead of their predators, they rapidly become the *plat du jour*. Such a biological 'arms race' between co-evolving predators and prey has been played out for millions of years on a four-dimensional stage in southern Australia: from the vast expanse of the Nullarbor Plain of Western Australia, east to the cliffs of the winding Murray River, to the spectacular coastline of Victoria. Over this large area thick sedimentary deposits were deposited in shallow seas between 45 and 10 million years ago. The rich fossil deposits are dominated by molluscs and sea-urchins, many of the urchins showing evidence of having fought a long, drawn-out battle with one particular group of molluscan predators over millions of years.

These days urchins suffer attacks from a wide variety of predators, including other urchins, starfishes, countless species of fishes, marine snails, crustaceans, birds, sea otters, arctic foxes, and of course, humans. Of these, only snails, which neatly cut or drill holes in the shell (known as the test), leave a recognisable trace in the fossil record. Living urchins, even burrowing urchins, like sand dollars and heart urchins, are known to suffer high levels of predation from a family of marine snails known as the Cassidae, or helmet shells. It is possible to identify cassids as the culprits by their distinctive method of penetrating the urchin's test, producing a neat, circular hole in the urchin's shell making it look as though as the animal has been shot. Other predatory snails, such as naticids, attack their prey by boring into the shell, leaving a characteristic incision which has bevelled edges. But cassids are much more clinical in their method of dispatching urchins. They cut a neat disc from the urchin's test, with the result that the edges of the hole are perpendicular to the surface of the test.

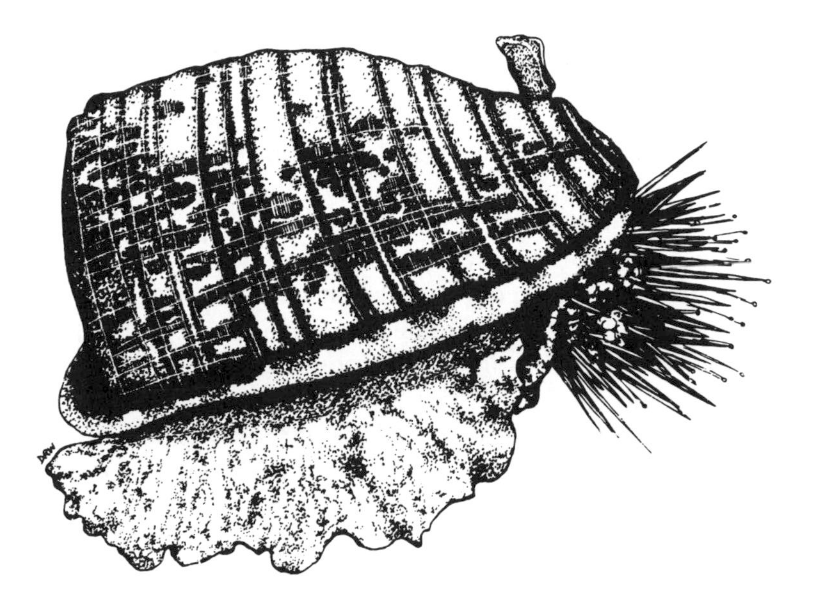

Come here my pretty! A *vampirish cassid snail spreads its shroud over a sea-urchin, giving it its fatal kiss of death. The defensive spines of the urchin are no match for the leathery foot of the killer snail. Increased predation by cassids on sea-urchins drove some to deeper, quieter environments to escape the onslaught. Others adopted the strategy of larger body sizes, setting up an evolutionary arms race with the snails.*

When they are hunting for urchins, the snail's siphon and tentacles protrude a little way beyond the shell. As they approach the urchin, the siphon and tentacles become fully extended. Once contact has been made with the urchin the snail raises its foot and arches over it, like a molluscan Nosferatu, before lowering itself over its victim. After having extended its proboscis, it begins its kiss of death. The tip of the proboscis is armed with a fearsome array of cutting teeth and is known as a radula. This is used to dissect a neat disc out of the urchin's test. The proboscis is then inserted through the hole, and the victim's slow death begins. This initial attack, after its phase of stalking, may have taken the snail only a few seconds. However, it feeds in a much more leisurely manner, taking up to an hour to eat its meal of fresh urchin gonad. Although there is little need to bolt its food, the initial rapid pinning of the urchin is critical, because in situations where an urchin has been observed to have been disturbed by a snail, but managed to escape the snail's less-than-amorous embrace, it has always been the urchin that outpaces the snail. In order to cope with an urchin's fearsome array of spines, cassids possess a tough foot that can push these spines aside. They also have the ability to track and locate urchins that are completely buried in the sediment, so long as they are not too deep.

The rich Cenozoic urchin fauna of Australia contains many individuals, as well as many species, that show the tell-tale signs of having suffered snail predation – neat, circular holes. Some fossils are known that bear healed, circular depressions. These are likely to represent failed predation attempts. The frequent occurrence in the Australian Cenozoic rocks of urchins bearing snail-inflicted incisions permits study of a number of aspects of the impact of predation pressure on urchins[1]. These include the effect that changing levels of predation had on the direction that evolution took, as well as the time of arrival of cassid snails in Australia. For there was once a time when urchins didn't have to cope with having (metaphorically speaking) to continually look over their shoulders, waiting for a cassid to pounce.

The first cassids are known from 40 million-year-old Late Eocene rocks in the Caribbean region. The first recorded evidence of their predation on urchins is on a specimen of the same age from Cuba. However, Australian Late Eocene urchins show no evidence of having been attacked by cassids, and neither have cassids of this age been found in Australia. But once they appeared in this part of the world 25 million years ago, they got stuck into the urchins straight away. Levels of predation on urchins in these Late Oligocene rocks in Victoria, such as on species of the heart urchins (spatangoids) *Lovenia* and *Eupatagus* were low, less than 10% having been killed by snails.

These urchins would have been shallow burrowers in the sediment. The holasteroid *Corystus dysasteroides*, which was probably not a burrower, although being heavily preyed upon in younger, Early Miocene strata, shows no evidence of lethal gastropod predation. However, another type of shallow burrowing urchin, the cassiduloid *Australanthus australiae*, shows higher levels of predation, with 37% of specimens bearing the neat circular incisions that are the fatal calling card of cassid snails. This variability in degree of predation is seen in other formations and points to differences in behaviour by different types of cassids and different food preferences for different types of urchins. It also indicates which urchins were more successful at being able to withstand the onslaught of this new predator.

The greatest levels of snail predation on the Australian Cenozoic urchins have been documented in 20 million-year-old Early Miocene rocks in South Australia that outcrop in the bright yellow cliffs of the Murray River[1]. These rocks, known as the Mannum Formation, contain rich urchin and cassid faunas. Of the more than 600 fossil urchin specimens that I have examined, one-third would seem to have been killed by these snails. However, there is a distinct variability in the proportions of individuals within different species that have been sucked to death by cassids.

The highest level was in the urchin that earlier in the Late Oligocene, was left unscathed – the holasteroid *Corystus dysasteroides*. In these younger rocks 60% of specimens have the tell-tale hole. The two most common genera in this fauna, the heart urchins *Lovenia* and *Eupatagus*, must have spent a good deal of their lives trying to avoid the clutches of cassids. A real war must have ranged beneath the sediment surface, for in the case of the three species of *Eupatagus* that I investigated, levels of predation were very high and very consistent, varying between 46 and 48%. In *Lovenia* the level was just below 30%. But where the holes are found on the fossils is anything but random, suggesting that the cassids knew their business, knowing just where on the urchin they should attack to get at their food most easily.

It is known that today there is a strong degree of host specificity between living cassids and particular species of urchins. Not only that, but many species of snails seemed to have a preference for attacking urchins in a particular spot. For instance *Cypraecassis testiculus* preferentially penetrates the large spiny, regular urchins *Tripneustes ventricosus* and *Echinometra lucunter* through the mouth, whereas *Cassis tuberosa* prefers to consume its prey through the side of the test. Likewise in the fossils. Not only are there differences between *Lovenia* and *Eupatagus* in where the holes are concentrated, but their distribution over the test differs between the three species of *Eupatagus*. In *Lovenia* more than half of the holes are on the upper surface of the test, a third on the side, and only 10% on the under surface. Many were concentrated around the anus. By contrast, the holes are more evenly distributed in the three species of *Eupatagus*. However, one species, *E. wrighti*, shows a preponderance of holes on the under surface (45% compared with 36% and 39% in the other two species). The other two species have fewer incisions but they are concentrated on the upper surface.

From this it would seem that, like today, 20 million years ago certain cassid species took a liking to certain urchin species. The influence of the primary spines in protecting urchins against snail predation is shown by the low frequency of incisions either in the region of the test covered by tubercles, or immediately posterior in the area covered by the canopy of extended spines, for both *Lovenia* and *Eupatagus* have fairly large, defensive spines on their upper surface. The high concentration of holes around the anus in *Lovenia* suggests the cassids had a predilection for this region, much like the living cassid *Cypraecassis testiculus* for *Diadema antillarum*.

There are other shallow burrowing urchins in the Mannum Formation. Nearly half of the the cassiduloids *Studeria* and *Notolampas* were killed by cassids. Even more spiny, surface dwelling urchins suffered quite high levels of predation. One, *Ortholophus*, shows 38% of specimens with incisions. However, some urchins seem to have escaped the attention of cassids. None of the small species (less than 10 mm long), such as the urchins of *Fibularia* and *Scutelloides*, show evidence of having suffered predation from snails. Neither do large species that exceed about 80 mm in test length, such as *Pericosmus* and another, large species of *Eupatagus*, show evidence of snail predation. This is consistent with the idea that snails adopt an 'optimum foraging strategy', concentrating only on prey of a certain size, consistent with their own food requirements. The larger the predator, usually the larger the prey. Conversely the smaller the predator, the smaller its prey. It seems likely, therefore, that only cassids of a certain range of sizes were present in these Early Miocene seas, so that any small or any particularly larger urchin species would not be preyed upon.

Arms races may develop between pairs of predators and prey, with each evolving ever larger species, the prey species trying to outgrow the predator, the predator then growing larger to cope with its larger prey. For example, modern species of cassids and *Lovenia* are both much larger than their Miocene counterparts. Australian Miocene cassids were no larger than 60 mm (about twice the size of their prey). Living species of *Lovenia* reach up to 120 mm in length, while coexisting cassids are at least twice this length. At the other end of the scale, if the potential prey species is too small it is not worth the predator's time in attacking it, as too much energy needs to be invested to make the efforts worthwhile.

There is another, major group of urchins that seems not to have suffered the slings and arrows of outrageous cassids – sand dollars. In the Mannum Formation these are represented by a species of *Monostychia*. I have only ever seen one that appears to have been killed by a gastropod. The reason for this resistance to the snail's unwanted advances may well lie in internal structures present within the sand dollar. Cassids feed by inserting their tube-like proboscis into the hole that they have neatly cut into the test, then locating and sucking out the gonads. However, sand dollars have a series of internal vertical rods and partitions, that are likely to have greatly frustrated the cassids' attempts to gain access to the urchins' gonads. Although selected primarily for other purposes, such as strengthening the very flattened test, it is possible that these internal structures made the sand dollars eminently preadapted to resisting the snail's attempt to the turn the sand dollars into their lunch.

At least that seems to have been the case relatively early in the history of this predator–prey relationship. In more recent times it looks as though cassids have managed to overcome this problem and, to some degree, outwitted the sand dollars in this continuing arms race. A species of the sand dollar *Peronella* that occurs in the 2 million-year-old Late Pliocene Roe Calcarenite of southern Western Australia shows high levels of snail-inflicted predation. More than half of these sand dollars appear to have been killed by cassids. However, unlike *Monostychia*, *Peronella* has fewer internal partitions. Consequently there would have been little in the way of barriers stopping the proboscis from reaching the urchin's gonads.

The extent to which different urchins were preyed upon varies not only with different types of urchins, but can vary within single lineages over time. The significance of this is that it provides an insight into just what might have been pushing certain animals to evolve in a particular direction. It has been possible to assess this using three species of the heart urchin *Lovenia*. In addition to the Early Miocene species (*Lovenia forbesi*), the Australian lineage contains a Middle Miocene species, called *Lovenia bagheerae*, and a Late Miocene species, *Lovenia woodsi*.

There are appreciable differences in the extent of predation in these three species. The oldest species suffered 29% lethal cassid predation. This dropped to 20% in the Middle Miocene species and to only 8% in the Late Miocene species. It is interesting that this youngest species shows a lower percentage of successful borings. Not only can the levels of predation be shown to have decreased between species with time, but this happened even within the oldest species, where the earlier forms experienced predation levels of 38%, but later forms dropped to just over 30%.

Predation upon different size-classes indicates that there was an optimum foraging strategy adopted by the predators. All three species

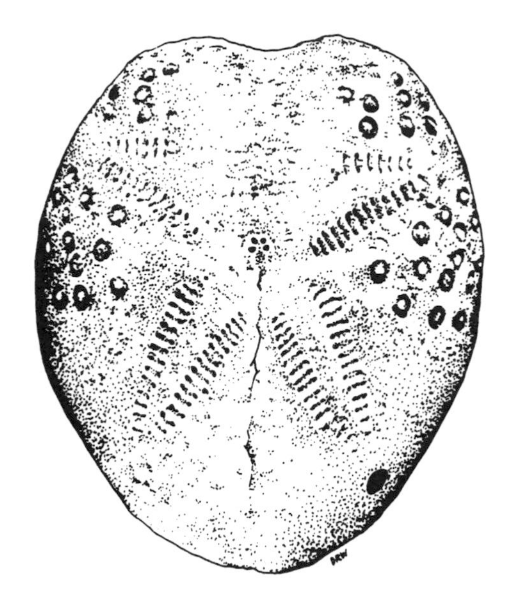

Ouch! The little hole on the bottom right side of this sea-urchin Lovenia, *represents a successful attack by a cassid snail that drilled through it and sucked out its gonads. What a way to go! Predation pressure has been a major driving force in evolution.*

show lower than average levels of gastropod predation in the smaller and larger size classes. Moreover, variable distribution of predation holes in the tests suggests that different cassid species were feeding on the different urchin species. So, for example, in the earliest species 56% of holes are on the upper surface of the test, with 34% in the side and only 10% on the lower surface. However, in the middle species 80% are on the upper surface, whereas in the youngest species, *L. woodsi*, there is a relatively high proportion on the lower surface.

So, why should the predation levels have dropped through time. Had the cassids lost the knack of dispatching as many urchins? Or had the urchins developed the ability to outwit the cassids? We cannot know just what behavioural strategies the urchins might have adopted to try to minimise the chances of being eaten, but we can interpret any changes in the environments in which they lived by analysing the sediments in which they are preserved. The reduction in the level of snail predation in the lineage is combined with the occupation of a finer-grained sediment by later species. This suggests that selection favoured those morphotypes that were able to occupy finer-grained sediments than their ancestors. They did this by being equipped with more burrowing spines, enabling them to burrow into finer-grained sediment. This is likely to have been in quieter, deeper water, suggesting that evolution has proceeded from shallow into deeper water.

There are major differences in the concentration and distribution of the defensive spines in the three species. Whereas the earliest species had a more widespread coverage, the descendent Middle Miocene species had a higher concentration of defensive spines on its upper surface, suggesting selection pressure initially favoured those animals with greater concentrations of spines – this has been termed a 'fortification strategy'[2]. Subsequently, selection pressure favoured those forms that appeared in the Late Miocene, which, although a little surprisingly had fewer defensive spines, had more burrowing spines on their under surface, making them capable of burrowing in finer sediment in deeper water. There is evidence from the living distribution of cassids that fewer of these urchin killers would have lived in deeper water, so any urchin able to colonise a deeper water environment would have a greater chance of surviving. This occupation of a region of lower predation pressure is known as 'habitat selection strategy'[2]. There would have been less selective pressure to produce a high density of defensive spines.

We see much the same positive correlation between a reduced defensive spine density and the habitation of deeper water in living species of *Lovenia*. Today two species of *Lovenia* live off the coast of northwest Australia: an inner shelf, shallow water inhabitant, *L. elongata*, and an outer shelf, deeper water inhabitant, *L. gregalis*. The shallow water inhabitant has twice the defensive spine concentration of the deeper water species. Indeed, of the eight described living species of *Lovenia*, the highest density of defensive spines is in species inhabiting water depths of less than 100 metres.

There is therefore strong evidence that predation pressure can channel evolution in certain directions. But this can be seen not only at the species level, but also at the level of families and orders, suggesting that the overall pattern of urchins, and indeed of many groups of organisms, may have been strongly influenced by predation pressure.

'Irregular' urchins are those that are not like mobile pin cushions, but have a covering of finer, hair-like spines and which either live on sediment or burrow in it. One of the major groups is the cassiduloids. These were common in the Mesozoic and in the early part of the Cenozoic, 50 or so million years ago. But with the evolution of sand dollars at this time, and their great evolutionary radiation, the cassiduloids went into a rapid decline in diversity. For instance in the Late

Eocene in Australia they comprise 25% of the fauna. But by the Early Miocene they were reduced to 10%. This niche displacement (for both occupy a similar shallow water, shallow burrowing mode of life) I suggest arose from different abilities of these two groups to cope with an invasion of a new killing machine, the cassids. The much lower levels of snail predation on sand dollars, like *Monostychia*, suggests that selection may have been favouring those forms that suffered less predation pressure. The same low levels of predation on sand dollars holds true in other parts of the world, where the mean level of cassid-induced deaths is slightly over 4%. If, as seems likely, the internal supports and rods restricted access by the cassid's predators to the urchin's gonads, then this can be interpreted as a 'maze strategy' of survival. The earliest sand dollars were tiny, little larger than a grain of rice. While many species grew very large, some, like *Fibularia* and *Echinocyamus*, remained tiny. By doing so, they can be considered to have adopted a form of 'crypsis strategy' by evolving a small body size that effectively took them out of the range of prey-size selection by cassid snails.

The dominant members of the Australian, and indeed many Cenozoic faunas around the world, were the heart urchins. Although some were subjected to high levels of predation, rapid evolution of new forms and adoption of a range of survival strategies meant that they have kept one lurch ahead of the ever-pursuing cassids. For example, the evolving *Lovenia* lineage, adopted the 'habitat selection strategy', where a number of different groups of heart urchins evolved into deeper water niches where they were subjected to less intense predation pressure. Another strategy was the 'fortification strategy', whereby there was an increase in density of defensive spines. Prior to the appearance of cassids, heart urchins did not have such large, defensive spines. Thus no Cretaceous or Paleocene heart urchins possessed primary spines, whereas of 123 genera recorded from the Eocene 25 had primary defensive spines.

The 'crypsis strategy', adopted by a number of heart urchins involved the evolution of morphologies that allowed them effectively to hide from predators by burrowing deeply into fine-grained sediment. Although cassids are known to burrow in search of their prey they are restricted to shallow burrowing in relatively coarse-grained sediments. Consequently, a range of morphological adaptions evolved in heart urchins to enable them to cope with living deeply buried in mud – anything for a quiet life. The last, and perhaps one of the most sophisticated strategies is the 'chemical warfare strategy'. Some species, such as the living heart urchin *Meoma ventricosa*, when attacked by a potential predator respond by secreting a yellowish pigment that is known to be toxic to other organisms[3]. Studies of this urchin revealed that relatively few specimens are ever attacked by cassids, probably on account of its method of chemical 'warfare'.

A feature of many heart urchin lineages, and indeed many fossil groups, is the frequent occurrence of patterns of speciation along an evolutionary gradient, in which ancestral species fail to persist after the evolution of a descendant, even though their ancestral habitat remains. The driving force behind such evolutionary trends may be predation pressure acting as the principal agent of selection. Those ancestral species incapable of withstanding high levels of predation become extinct. Descendent species will persist only for as long as the levels of predation remain below a certain critical threshold. As predation pressure increases, only peripheral isolates with more effective anti-predation strategies will persist, and provide the seed for the evolution of a new species[4].

Large-scale evolutionary trends in sea-urchins reflect many of the observed interspecific-level shallow to deeper water evolutionary trends induced by predation. Dave Jablonski of the University of Chicago and Dave Bottjer of the University of Southern California, analysed onshore to offshore patterns in 13 orders of post-Palaeozoic sea-urchins[5]. They discovered that seven of these show a trend of migration from onshore to offshore environments. Migration offshore occurred in two ways. One was by an expansion of some forms of a lineage into deeper water, with others remaining in shallow water, so increasing the degree of ecological diversity. The other was by a displacement, whereby there was a loss of onshore representatives, descendants evolving into a deep water habitat. This is shown by certain forms being present in shallow water deposits in, say, Early Cenozoic deposits, but today occurring only in relatively deep water. One example of this is the Holasteroida, a group of primitive heart urchins that evolved in the Mesozoic, and went into a steep decline through the Cenozoic. One form, that I have mentioned earlier, *Corystus*, is quite a common element in shallow water deposits in Eocene to Miocene strata in southern Australia. The only member of the Corystidae remaining today lives in deeper water on the outer shelf. Similarly a little heart urchin present in the same, shallow water deposits, *Hemiaster*, today lives only in water thousands of metres deep.

Steve Stanley of Johns Hopkins University, Baltimore, has suggested that the disappearance of Palaeozoic urchins, which possessed flexible tests, and their replacement by urchins with rigid tests, may have been partly due to predation pressure[6]. Similarly, the great increase in diversity of infaunal urchins, snails and bivalves since post-Palaeozoic times has been shown to correspond with an increase in diversity of predatory snails. The general pattern of urchin evolution, from forms that grazed on rocks, to forms that could burrow shallowly in coarse sediments, to those that could inhabit deep burrows in fine sediment in deep water, probably has its ultimate cause in increasing levels of predation.

While I have talked at length about how predation has directed urchin evolution, many palaeontologists have interpreted the patterns of changes that they see in the fossil record in many groups of organisms in terms of the influence of changing levels of predation. For instance, Steve Donovan of the University of the West Indies and Andy Gale of the University of Greenwich, London, have argued that the blame for the decline of articulate brachiopods after the great Permo-Triassic mass extinction event can be laid firmly at the feet of predatory starfish[7]. While they first appeared some 450 million years ago in the Ordovician, starfishes underwent a major increase in diversity at the beginning of the Mesozoic. Significantly these starfish evolved characters not present in the pre-Mesozoic forms: muscular arms, suckered tube feet, flexible mouth frame and reversible stomach, all key elements in a predatory mode of life.

Unlike urchins, brachiopods failed to evolve many anti-predation strategies. A few tried the chemical warfare technique in producing substances that gave them a bad taste, while a few evolved a cryptic habit. Bivalves, on the other hand, another likely source of food for a peckish starfish, have radiated successfully in post-Palaeozoic times. They have evolved far more anti-predation strategies, such as hiding, by burrowing in the sediment, swimming, in the case of scallops, ability to live exposed above the water level at low tide, and growth to a very large size. The only way to survive the evolutionary arms race is to come up with a strategy to defeat your attacker or, when the boot is on the other foot, to evolve more efficient means of dispatching your prey. To eat, or to avoid being eaten, is a potent symbol of the driving force of evolution.

17

Children of the Evolution

Human evolution from the inside

The next time that you take a trip to the zoo, try and drag yourself away from the lions, tigers and giant pandas, and go and have a look at the chimps. In particular look closely at the baby chimps. You have to admit that a tiny, new-born chimp is very cute – those big, doleful eyes and delicate face nestling beneath a relatively big skull. Contrast that with its mother and father – anything but cute – protuberant face and massive jaws set beneath a relatively small skull. Like humans, chimps change in appearance a lot as they grow.

When you go home, look in the mirror and ask yourself the question – 'Do I look more like the baby chimps, or their parents?' Chances are that most of us (I hope) would see a much closer resemblance to the baby chimps, for like them we too have a relatively large skull, small face and jaws, compared with other adult primates. So, what does this tell us, if anything, about how we, as a species, might have evolved? If the chimps have more in common with our shared ancestor than they do with us, this would seem to suggest that the apparent retention by us of a baby primate face may be reflecting the underlying processes that controlled our evolutionary destiny. But can our tortured path through millions of years of evolution really have led us to end up as little more than baby apes that have failed to grow up? For that is the inescapable conclusion reached by many anthropologists and biologists. However, another group of evolutionary biologists, concerned with the relationship between development and evolution, have argued that such apparent similarities are very misleading. Our evolutionary history has followed quite the opposite pathway, leading us to develop both morphologically and behaviourally beyond the apes. Yet our reflection in the mirror still whispers 'baby ape'. Herein lies the paradox of the baby-faced super-ape.

Deep inside each of the cells of our body lies a tiny clock that controls the beat of life. These little timepieces determine the rate at which our cells divide, the rate at which we grow and develop, when we are born, when we mature and when we stop growing. If the clocks slow down, or if they speed up, the effect on evolution can be quite profound. Recently, the debate has resurfaced on how fast the clocks tick in humans. Have they, as many have argued, slowed down as we have evolved, resulting in us looking like our ancestor's children? Or have some ticked faster, or maybe for longer, and turned us into the ultimate super-ape?

Much of the debate on human origins has centred on the timing of the morphological changes in hominid evolution that have taken place over the last 4 million years or so – when we achieved bipedalism; the sequence of changes to our cranial anatomy, in particular our brain size; and all the adaptational consequences of such changes. While studies of evolution in general have focused for decades on genetics and natural selection, there has been something fundamental missing from the evolutionary equation. This missing link is an understanding of how any genetic changes manifest themselves to produce changes in shape and size – the fodder upon which natural selection can feed. After all, it has often been remarked that we are about 99% genetically identical to chimps – yet we look and act so very differently. Why?

Self-evident as it may be, evolution is all about time – but not just on the scale of thousands or millions of years. It also involves the subtle changes to the timing of our cellular clocks, and the effect of this on the extent to which we, as individuals, change and develop during our lifetimes – our developmental histories. As we and other animals have evolved, the shapes and sizes of all our body parts – our head, our arms, our legs, in fact everything – will have experienced either more or less change than our ancestor during development. Our clocks may have ticked slower or for a shorter time. Then again, they may have ticked faster or for longer, taking our development down new and uncharted pathways, leading to new shapes and sizes. If a descendant undergoes less change during its development it will retain ancestral juvenile features – this is known as *paedomorphosis*, literally 'child-like shape'.

As we have seen in Chapter 7, this played an important role in insect evolution. If a descendant undergoes more change, and development continues 'beyond' that of the ancestor, such as in many dinosaurs (Chapter 14) and whales (Chapter 15), it is called *peramorphosis*, literally 'beyond shape'[1].

As individuals we each have our own destiny – our life history, extending from the magic moment of our conception, when life begins, until it ceases when we die. As we are catapulted through our embryonic, then post-embryonic, development, until we stop growing as we pass into adulthood, we grow up. 'Ho, hum,' I hear you say. But think about it. We do this not only by getting bigger, but also by changing shape, particularly the relative shapes of different parts of our bodies. This shape change is very dramatic during our life as embryos, less so as infants, and still less as we march through our childhood and become adults. For instance, the head of a newborn baby is much larger relative to the size of its body and the size of its limbs, than it is in an adult, for it has grown at a phenomenal rate in the womb, compared with other parts of the body. As the baby develops out of the womb, the time comes for its limbs and trunk to catch up and increase in size and change in shape more than the skull. Even so, the skull (and so also brain volume) still increases during childhood, but less markedly than earlier. By the end of our first nine months of existence, our brain has reached 25% of its adult volume. It then takes another five years to reach 90%, then a further five years to get to 95%.

The key to how animals evolve different shapes and sizes lies in heterochrony – the tinkering with these relative shape changes, by increasing or decreasing the extent of growth of different parts of the body, compared with the ancestor – changing how fast the clock ticks and for how long. Natural selection then sifts through the various permutations and, depending on a great constellation of environmental conditions existing at the time, favours certain shapes and sizes. So, as a mammal, for example, grows and develops, it may experience greater change in limb size and shape than its ancestor, but other parts of its body may grow just the same as in the ancestor. In some cases, some parts may even develop less than in the ancestor. For most species are a cocktail of paedomorphic and peramorphic features. Some, like *Tyrannosaurus rex* and many of the larger whales, are spectacularly potent cocktails. But what about us? Are we, as some have argued, little more than juvenilised apes? Or are we peramorphic, and developed beyond ancestral apes? This may smack of nineteenth century ideas of recapitulation. Ernst Haeckel's view was that 'ontogeny recapitulates phylogeny', in other words development is a reflection of our evolutionary history, and descendants have gone through more changes than their ancestors. Yet, this perspective shouldn't be summarily dismissed. Alternatively, could we be a mixture of the two, like so many other animals?

For as long as anthropologists and biologists have pondered over the role of heterochrony in human evolution, our view has been coloured by the 'baby ape paradox' – that our evolution has been entirely by paedomorphosis. Great lists have been drawn up to support this notion, listing features from small teeth to the form of our external ear. Intuitively it seems slightly ridiculous that our evolutionary success has been due to our being little more than overgrown baby apes. Perhaps we feel slighted by the thought that our perceived position, perched on top of the evolutionary tree, has come about from our being little more than failed apes. During the last few years there has emerged a realisation that we are, indeed, much more than this, and that our evolution has been driven from within by quite the opposite effect. Our internal clocks have not slowed down – they have, to a large degree, either sped up or ticked for a much longer time[1].

First we need to solve this 'baby ape paradox'. Brian Shea of Northwestern University in Illinois has argued that we appear to be stuck with the head of a juvenile ape because the extent to which each part of our head grows during development, relative to the other, varies between us and other primate species[2]. The shape of a chimp's face changes more than ours as it grows up, becoming, along with the lower jaw, more prolonged. By contrast, the skull in humans undergoes a spectacular increase in its growth during development, whereas growth of the face and lower jaw fail to keep up with this. Compared with other primates, including our extinct ancestors, modern human patterns of facial growth are a direct result of unique growth patterns involving, in part, resorption of bone over much of the midface. So, far from our head showing entirely paedomorphic features, the skull (and thus the brain) undergoes far greater development and increase in complexity more than in any other primate – this feature is peramorphic, having developed 'beyond' the ancestral condition.

Despite this, it could be argued that some of our anatomical features are reminiscent of juvenile apes: our general lack of long body hair, our thinner skull bones, smaller teeth, relatively shorter arms and toes. But these really are insignificant when compared with the major anatomical changes that have occurred throughout hominid evolution. Body weight and size have more than doubled, brain volume more than trebled, and the hindlimbs lengthened relative to our trunk. Such changes contributed greatly to our evolutionary success, and gave rise to our bipedal gait and our large cerebral cortex – the seat of our conscious thought, of our memory, our intelligence, and our speech. These are not the features of a baby ape. On the contrary, we have developed beyond our ancestors in these traits, and well beyond other primates.

The key to the underlying changes to development that fashioned our evolutionary history lies not in how fast or how slow we grow, but for how long. For herein lies the answer to our evolutionary success. Compared with all other primates we live for a very long time. Our 'three score years and ten' exceeds all other primates. But this prolongation of our life occurred not simply by extending our time spent as adults. What is special about human evolution is that all growth phases are very long – each has been stretched out.

'Perhaps the most crowning moment I have ever spent as a palaeontologist was the day in late 1994 when I held the Tuang skull in my bare hands.' These are the words of my co-author John Long, and sum up the impact that a few, special fossils can have. Amongst these are a few of the precious hominid fossils that the Earth has yielded up over the last 70 years or so since Raymond Dart described this first australopithecine skull in 1925 – the Tuang skull. Australopithecines were the earliest hominids, in other words they were apes that could walk upright. They differed from the later species of *Homo* in being smaller, having smaller brains, but relatively larger jaws and teeth. What these fossils tell us is that our very early ancestors were australopithecine primates who lived up to 4.5 million years ago and were much smaller than us. What was so spectacular about Dart's australopithecine child's skull was that a 2.5 million-year-old primate skull had been found in which the hole through which the spinal cord ran was at the base of the skull. This indicated that the child walked upright. Dart called the skull *Australopithecus africanus*. Since then even older australopithecines have been found. The most well-known is 'Lucy', the incomplete 3.2 million-year-old skeleton of another species, *Australopithecus afarensis*, found by Donald Johansen and his team in 1974 in Hadar in Ethiopia. As the oldest hominid known at that time, Lucy was seen as being close to the link between apes and humans. Lucy and these other australopithecines were about the size of a chimpanzee. Their brain was much smaller than in later hominids, and the arms were relatively longer, whereas the legs were relatively shorter. As we shall see, of some significance was the shape of their thorax. It was rather ape-like in being cone-shaped, indicating the possession of a large gut, a prerequisite for a predominantly plant-eating primate.

Northeast of Lake Turkana in Ethiopia, at a place called Aramis, the oldest australopithecine to date has been found. Tim White of the University of California at Berkeley and his colleagues described 17 bones from a hominid that lived between 4.3 and 4.5 million years ago[3]. In 1994 another mandible and a partial postcranial skeleton were found. So different are these remains from other australopithecines that they have been called *Ardipithecus* (meaning 'ground ape'). Compared with *Australopithecus*, *Ardipithecus* has relatively larger canine teeth, which, along with the molars, have thinner enamel, plus differences in the premolars. Small, like other australopithecines, *Ardipithecus* would have weighed in at, perhaps, about 30 kg. Other fossils found at the site by White and his colleagues include the remains of monkeys and a large bovid, the kudu. Because both of these are forest inhabitants today, a similar mode of life is envisaged for *Ardipithecus*. This calls into question the premise that hominids evolved in the open plains. Quite possibly bipedalism arose in the forests. As the forests shrunk while the world dried out as it sped into the last great Ice Age, such apes would have been preadapted for life on the savanna.

There have been countless suggestions as to what drove one small group of primates to achieve a bipedal gait and be able to walk upright. Some explanations smack very much of Lamarckism, such as the proposition that prehominids stood on two legs to reach fruit hanging down temptingly from the branches of trees above their heads. Others think that it was a means of keeping the brain cool. Stand upright, the argument goes, and you reduce the chances of sunstroke (well, if you lie on the beach, you stand more chance of getting an overall suntan, than if you stand up . . .). A more recent suggestion, made by Nina Jablonski and George Chaplin of the California Academy of Sciences, offers an intuitively more satisfying explanation related to bipedal threat display. They argue that selection strongly favoured those individuals who could stand taller during the frequent threat displays between individuals. The longer (and the higher) an individual could stand, the more successful it would be in the troop's hierarchy, and the greater the chance of mating[4].

Australopithecus afarensis lived between 3 and nearly 4 million years ago. At 1.7 metres tall, and weighing between 30 and 45 kg, it was a little larger than both *Ardipithecus ramidus* and *Australopithecus*

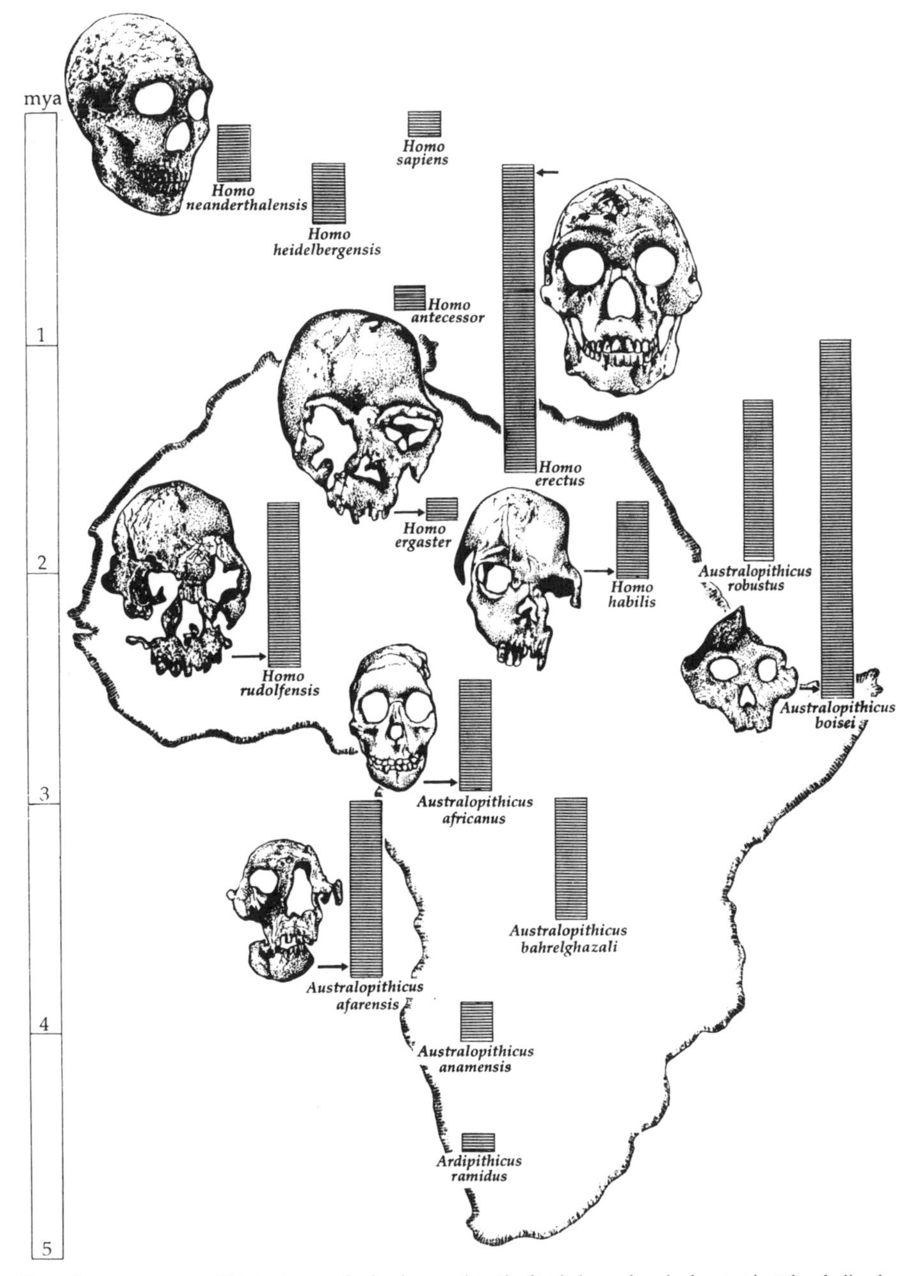

Get a brain and get a life! Africa in the background is the birthplace of early hominids. The skulls show the progress of forms, increasing in cranial capacity with time (dates of occurrence indicated by bar graphs), as humankind, beginning with the genus Homo, *emerges from the* Australopithecus *lineages some 2.4 million years ago. The position of the skulls on the map is not meant to be indicative of where they were found.*

africanus. Its brain was much smaller than ours, with a volume of about 400 cubic centimetres (cc), compared with ours at nearly 1400 cc. However, does it mean that just because these early primates were smaller than us they matured at an earlier age, or were their growth rates less than ours? One way to determine this is to look at the relative times of eruption of teeth. From these we can deduce the relative lengths of different growth phases.

Using the age at which teeth erupt, hominid post-natal growth can be classified into three phases: infantile, juvenile and adult, based on periods before, during and after eruption of permanent teeth, respectively. Timing of tooth eruption, as Holly Smith of the University of Michigan at Ann Arbor has pointed out, tells us a lot about an animal's life history, such as its gestation period, timing of onset of sexual maturity and length of its life[5]. So, for instance, the later that permanent teeth erupt, the longer the gestation period will be and the later the weaning; consequently, the longer the infantile and juvenile dependency. The onset of sexual maturity will be later, and the body and brain may grow to a larger size. The lifespan will also be longer.

Teeth lay down daily markings in the enamel and seven to nine day cycles can also be identified. This permits an individual's age at death to be worked out. The time of eruption of the first molar tooth can also be reasonably accurately estimated from the brain weight, as there is a very high correlation between the two. The smallest living primate, *Cheirogaleus medius*, has a tiny brain that weighs only 180 grams. It also has the earliest emergence time of the first molar of all primates. Moreover, it matures very early, is very small and has a very short lifespan. At the opposite extreme is *Homo sapiens*, possessing the largest brain of any primate. We also take the longest time to erupt our first molars, have the longest childhood, and so the largest body and longest lifespan. With such correlations it is possible to estimate the life history patterns of extinct hominids.

Australopithecine first molars are thought to have erupted when the primates were between three and three and a half years old, and these hominids probably lived until they were 35 to 40 years old. In *Homo habilis*, which had a brain volume of 580–750 cc, and appeared about 2 million years ago, and early forms of *Homo erectus* that lived about 1 million years ago, the first molar erupted a little later, between four and four and a half years of age. These species may have lived until they were about 50 years old. In later *Homo erectus* (which had a brain volume of 900–1100 cc) and lived a few hundred thousand years ago, the first molars probably erupted at about five and a half years of age. The first molars of 'modern' humans, who appeared only about 120 000 years ago, erupt after six years. Our lifespan is in excess of 70 years and our brain volume about 1400 cc.

So what we are seeing with hominid evolution is a stretching out of all our growth phases – gestation periods were probably longer, the infantile to juvenile transition took longer to come about, and the transition to adolescence and adulthood took progressively longer as the end of childhood was prolonged. The extension in successive species of faster embryonic, infantile and juvenile growth rates in certain structures, principally the skull and brain, the lower limbs and foot bones, as well as body size as a whole, resulted in these characteristics developing 'beyond' the comparative stage in the ancestors. This longer period of development gave us our longer legs to carry our upright, large body, larger foot bones to support our greater weight, not to mention our large brain that enables us to ponder such questions in the first place.

Such dragging out of hominid life histories would also have had profound impacts on behaviour, social structure and diet. Dental evidence suggests that australopithecines were exclusively vegetarian, unlike later hominids that incorporated more animal products into their diet. This change in feeding habits may, paradoxically, be related to the increase in brain size. Growing brain tissue is metabolically a very expensive exercise. Despite weighing only about 2% of the total body weight, the brain uses about 17% of the body's energy. As a more than three-fold increase in brain volume occurred during hominid evolution, the energy to produce more brain tissue had to come from somewhere. One possibility is a developmental trade-off, the brain increasing at the expense of other structures – but which ones?

An intriguing hypothesis has recently been proposed by Leslie Aiello of University College in London and Peter Wheeler of Liverpool John Moores University in England, that may shed some light on this question. They have suggested that the increased brain size occurred at the expense of another metabolically expensive organ – the gastro-intestinal tract – with bigger brains traded for smaller guts[6]. Compared with other primates our guts are quite small, being 60% of what they should be, in terms of the size of our body. Our brains, on the other hand, are larger than they should be. As gut size is correlated very strongly with diet, small guts only function with high-quality, easily digested food such as meat. To process large amounts of lower nutritional vegetative matter, large guts are needed. So, if selection has favoured our large brains, then a major developmental trade-off might have been with the gut.

Evidence for gut reduction during hominid evolution comes from changing skeletal shape. The rib cage of *Australopithecus afarensis* is funnel-shaped, being wider at the base than at the top, to accommodate a larger gut. Early hominids, like *Homo ergaster*, had a more barrel-shaped rib cage, suggesting a smaller gut and a different diet. Significantly, this change in diet was accompanied by the development of the use of tools that were probably mainly used in butchering. To get a bigger brain a more protein-rich diet was needed. The effect of the bigger brain was the attainment of increased behavioural sophistication that allowed this dietary change to be achieved. This involved the acquisition of the ability to use tools as well as the evolution of more complex social behaviours. With the reduction in gut came a reduction in jaw and tooth size, as hominids ate food that did not need endless chewing.

This appearance of species of *Homo* occurred about 2.5 million years ago. With it came a blossoming of species that were appreciably larger than the australopithecines, with distinctly larger brains, in addition to the differences in shape of the thorax. Much recent research into hominid evolution has centred on this evolutionary radiation of early species of *Homo*, and trying to understand their relationships to each other. The nice, simple picture of a straight evolutionary line from *Homo habilis* to *H. erectus* into *H. sapiens* has been replaced by a branching picture, in which early species like *H. ergaster* and *H. rudolfensis* formed part of the early branching along with *H. erectus* and *H. habilis*. From out of these early species of *Homo* marched *H. heidelbergensis* and *H. neanderthalensis*, and finally us, *H. sapiens*.

The branches of this tree became an even more complex maze in mid-1997, with the publication of the description of yet another new species of *Homo*, this time from northern Spain. A team led by José Bermúdez de Castro from the National Museum of Natural Sciences in Madrid uncovered the remains of six individuals from a Pleistocene cave site near Burgos in northern Spain. They call this species *H. antecessor*[7]. What is striking about this material is that the individuals seem to possess a mixture of very modern features, combined with some more primitive ones. Thus the midfacial morphology is very much like that of modern humans as shown by the face of one of the specimens, an 11 year old individual. They suggest that this juvenile feature has been

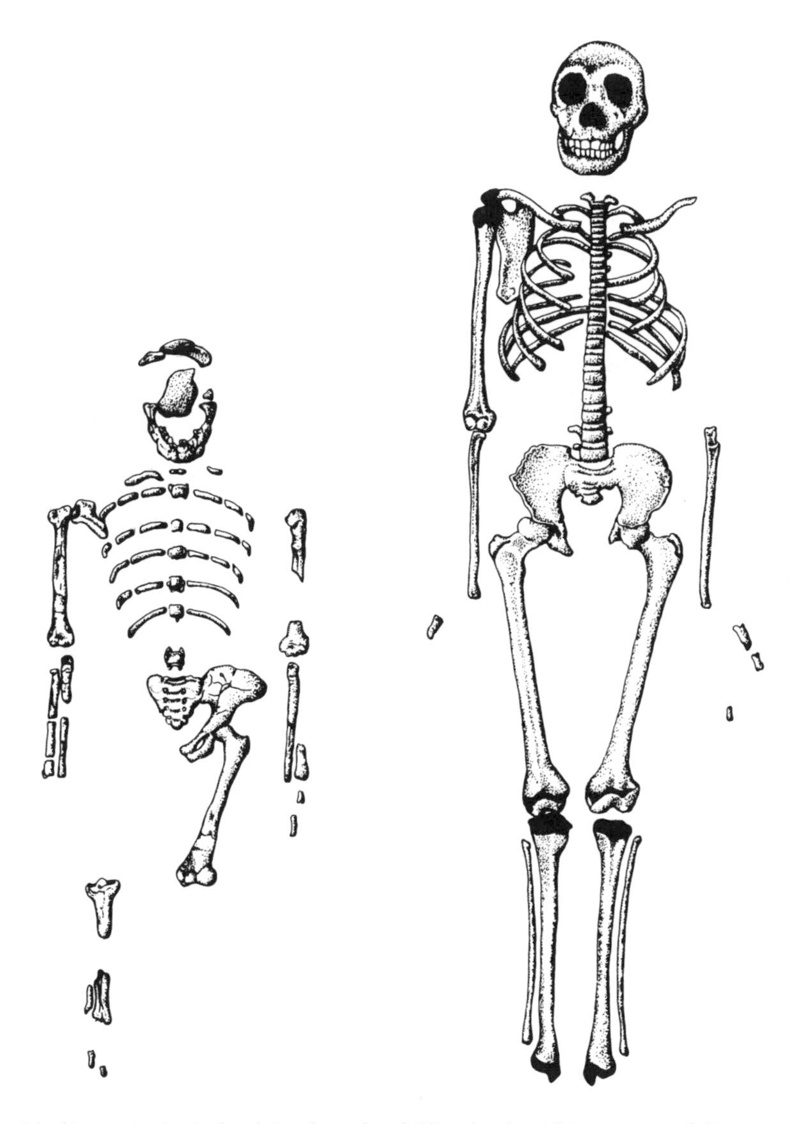

Lucy with this guy in time's hand. Turkana boy holding hands with Lucy, a symbolic gesture of the merging gene pools from australopithecines to early hominids. The little one is Lucy, an Australopithecus; *the big one, Turkana boy, is a* Homo.

retained into the adults of later species like our own. In terms of its more ancestral features, they argue that it has more in common with *H. ergaster* than it does with *H. erectus*. Consequently, de Castro and his colleagues have raised the possibility that forms like *H. antecessor* are on a more direct line between *H. ergaster* and *H. sapiens*, with *H.*

heidelbergensis and *H. neanderthalensis* being a side branch. Another side branch would be the *H. erectus* lineage.

However, irrespective of the exact details of which species evolved into which, the overall pattern is still there of sequential delays in transition from one growth phase to another resulting in the evolution of a larger, more complex brain through the 4.5 million years of hominid evolution. The consequence is that our cognitive capabilities are much more complex than in other primates. Extending the time spent as infants and juveniles produces an even more complex brain, because it is during these periods that dendritic growth of the neurons occurs – the longer the period of growth, the more neurons. Humans have easily the greatest number of interconnecting neurons amongst all primates. Compared with all other primates, living or extinct, we have a larger and more complex brain at the end of each growth phase.

Myelination of neurons (which allows information to be transmitted as much as 20 times faster than in non-myelinated neurons), along with greater development of synapses, are important as they are critical in maturing memory, intelligence and language skills. In rhesus monkeys myelination occurs for the first 3.5 years and then stops. In humans it continues well into adolescence. As well, dendritic growth continues for longer in humans than in any other primates, until about 20 years of age. Kathleen Gibson of the University of Texas at Houston has argued that neurologically the adult human brain is not paedomorphic. In every way it is peramorphic in having evolved well beyond that of all other primates[8].

While this overall 4.5 million year pattern is one of increased brain size, when looked at on smaller time scales there are periods when brain size may not have increased substantially for a certain period or may even have been reduced. Recent work[9] on detailed changes in body and brain size in Pleistocene species of *Homo* has revealed some surprising changes. Assessing body mass from a few scraps of bones and teeth has always been a problem. What the new research has revealed is that Pleistocene species of *Homo* were up to 10% larger than modern humans. The main increase in brain size seems to have taken place during the Middle Pleistocene, 600 000 to 150 000 years ago. Prior to this brain size had remained reasonably stable for about 1.5 million years.

The methods used to calculate these changes were measurement of the head of the femur, as there seems to be a strong correlation between this measurement and body mass. The other measurement they made was of the maximum width of the pelvis. Both measures have reduced in the last 50 000 years. Thus modern humans, compared with their ancestors, seem to have a smaller body mass, but a relatively larger brain size. However, in real terms the brain actually seems to have decreased in size a little, if these data are to be believed.

Be that as it may, one of the most important aspects of hominid evolution on the time scale of millions of years is the stretched-out growth period, and with it our longer childhood – much longer than in any other primate. This is important because it is during this time that we learn most readily. Our larger, more complexly interconnected neocortex is able to store greater amounts of information and to undertake more complex mental functions, such as processing and articulating complex language, manufacturing intricate tools and undertaking complex social development.

Like morphology, behaviour also changes during development – it also evolves. During our first two years of life, we discover the world around us – the properties of objects, space and time. Within this period our behaviour increases rapidly in complexity. For example, from two to four months of age simple repeated actions are carried out, like sucking the thumb. From three to eight months actions are repeated more often and may induce a reaction, like shaking a rattle and it makes a noise. Later, from, say, twelve to eighteen months, such actions are more complicated and involve cause and effect, like seeing how far food can be thrown across the room. From two to six years, imitative behaviour increases in complexity and unfolds into symbolic play, and imitative drawing and speaking. From six to twelve years of age concepts, such as of the conservation of quantity, weight and volume, become embedded. Finally, hypothetical deductive systems become established along with ways of looking at problems from different angles[10].

A leading worker in the field of the evolution of primate behaviour, Sue Parker of Sonoma State University in California, has pointed out that when compared with humans other primates do not develop behaviourally to the same extent. For example, great ape adults have a range of cognitive abilities similar to those found in two to four year old humans. Adult cebus monkeys have cognitive abilities like those of a two year old human child. Adult macaques pass through even fewer stages of development, showing cognitive abilities comparable to those of a one year old child. So, humans' protracted childhood leads to greater behavioural complexity.

Understanding the evolution of behaviour in our ancestors is obviously much more difficult. However, archaeological evidence, in the form of complexity of tools, provides indirect evidence to support the idea that the dominant trend in hominid behavioural evolution has been one of increasingly more complex cognitive development, as successive hominid species had longer and longer childhoods[10].

Early species of *Homo*, such as *Homo habilis*, manufactured very simple stone tools. However, the degree of stone tool complexity changed little for the next 2 million years, apart from the development of Acheulian hand-axes by *Homo erectus*. While complexity increased at the beginning of Middle Palaeolithic times, 125 000 years ago, as *Homo erectus* was replaced by early *Homo sapiens*, the great increase in sophistication of tools did not occur until Late Palaeolithic times, 30 000 to 35 000 years ago. This coincided with the blossoming of art and the evolution of complex cultural patterns of modern *Homo sapiens*.

Drawing on such evidence, Sue Parker and Kathleen Gibson have argued that cognitive development in adult australopithecines was equivalent to that of two to three year old human infants; *Homo habilis* like a five to six year old; *Homo erectus* a six to eight year old; and *Homo heidelbergensis* and *Homo neanderthalensis* a ten to twelve year old[10]. Changes in body size, brain weight, timing of tooth eruption, and maturation times, all suggest that the evolution of cognitive abilities in hominids has been one of a steady stretching out of mental and intellectual capabilities, along with the extended physical development. *Homo habilis* obtained its food using a simple worked stone. *Homo erectus* used a slightly more complex, worked stone. Early *Homo sapiens* crafted more intricate stone tools, or used other materials like bone. Late *Homo sapiens* takes a drive to the nearest McDonald's.

Like many other organisms, we are an evolutionary cocktail – some features mere distant echoes of our long-extinct ancestors' children; others, structures or behaviours that have evolved beyond anything in our ancestors' repertoires. This evolutionary 'give-and-take' has led to those features that characterise humans – long life, protracted childhood, large brains, large bodies and our bipedal gait. Baby-faced we may be – super-ape we certainly are.

18

Old Friends Revisited

A new look at classic examples of evolution

It's another cloudy, windswept day. The wet streets are trying their utmost to glisten as much as they can on such a dull, drab afternoon. You quicken your pace as the rain starts coming down harder. Just a few hundred metres to go and then you will be there. A car speeds past, close to the pavement, and sends a spray of muddy water over you. Will it really be worth it? But then, what else to do on such a foul day but visit the Museum. After all it's one of the few places of free entertainment left. Entertainment? Not much in this one. The Museum in this small country town has seen better days.

You round the corner and there it is – the red, brick mausoleum, resting place of so many dead bodies, and decaying feelings of the past. But you rush up the steps to get out of this persistent, nagging rain. The difference is stunning. The roar and hiss of passing vehicles, the clatter and the noise of the town are suddenly gone, as if they were just a figment of your imagination. Here all is dry; all is dust. The only sound is the dripping of water off your coat on to the cold floor, and the buzzing of the lamps. The dull, yellow lights spread a pale, sickly pallor over the stuffed, moth-eaten bear that greets you. The smell is the smell of museums. There is no other way to describe it. But not one to bottle and cherish. Once smelt it's never forgotten.

In to the small room to your left, past the flint tools, lovingly crafted by some distant soul too many thousands of years ago to even contemplate. Past the brown, chipped pots and the faded images of cavemen huddled round a fire. And then you find it. Tucked away at the back of the room, in a glass case that wouldn't object to a bit of sprucing up. There it is – the icon of evolution – plaster casts of bare bones from creatures said to be horses. But these were strange horses, sporting more than one toe and big enough only for a vervet monkey to ride on. How many museums around the world have not, at one time or another, had just such a case. Thankfully many are now gone, but some still remain, purporting to show the classic example of evolution of the horses – the epitome of evolution. A simple evolutionary trend from a four-toed, tiny horse to the massive equine thoroughbred that today imperiously snorts at you from its exalted height.

Usually labelled as *'Eohippus'*, this first ancient horse, that scurried through the woodlands of Western Europe and North America 55 million years ago, is now called *Hyracotherium*. The story tells of how this horse evolved into *Mesohippus*, then into *Mercyhippus*, then *Pliohippus* in more recent times, before being replaced by the mighty *Equus*, winner of Grand Nationals. All this in a little over 50 million years. And as the horses got bigger, their teeth became higher and more complex, allowing the later larger horses to move out into the newly evolving green pastures, to graze, rather than to browse, and to walk not on four toes, but on just one big hoof.

Lovely story. The trouble is that it's basically wrong. Evolution of such a major group is not like an arrow hurtling through time. It is a complex branch with many dead ends, and hopeless causes, and success and failures, and being in the right place at the right time, and avoiding being eaten, or becoming extinct. If you survive these hurdles you might be one of the winners in the evolutionary handicap. Not only that, but we often view such evolutionary changes that did occur purely in terms of what adaptive advantage it gave to the descendants, without trying to unravel the underlying processes that actually caused these changes to occur.

It has probably not escaped your notice that we have been mentioning in various chapters a process called 'heterochrony'. It's really an awful name, but unravelling its classic etymological pedigree tells you that it means, simply, 'changing time'. For so long our perception of how organisms change through time has been blinkered by a concentration on looking purely at adult characters. You know the sort of illustration, of humans evolving from some hairy ape into a dapper businessman, or of horses evolving from tiny animals the size of a fox to a massive shire horse. It is always depicted in terms of changes to the adult. Yet natural selection is not that picky. It is weaving its spell from the moment of conception, picking and choosing between different organisms, or parts of an organism as it undergoes its embryonic or larval development, through its childhood until it is an adult. In some animals, like lamprey fishes and some sea-urchins, there can be major differences in morphology and behaviour in the preadults of closely related species, but the adults look very similar.

Evolution is all about tinkering with what is already there. Massive genetic mutations are not necessary in order to evolve new shapes, new sizes and new behaviours. What is important in evolution are changes to the rate and timing of development – how fast an organism grows during its development, compared with its ancestor, and for how long. Like all scientific disciplines, this study of evolution operating from the inside has sprouted its own terminology. And what a terminology. So many words have been introduced that rather than clarifying what is going on it has led to confusion and general dismissal of the importance of the process.

Essentially all that is happening in heterochrony is growth and development passing through more or less stages than in the ancestor. If fewer stages are passed through it is called 'paedomorphosis' – meaning 'child shape'. If it goes through more it is called 'peramorphosis', meaning 'beyond shape'. Paedomorphosis can arise by growing at a slower rate, finishing growth earlier (for example, sexual maturity might arrive earlier) or starting growth later. Peramorphosis, on the other hand, can arise from faster growth, or from growing for a longer period, by delaying the offset of growth or turning it on earlier. While some species may show just one type or another, being either paedomorphic or peramorphic, most, like humans (Chapter 17) are a mosaic of both features.

Well, what about horses? One of the undoubted observations is that early horses were much smaller than modern horses. There is no escaping the fact. But rather than a progressive increase in body size over more than 50 million years, what happened was an increase in the diversity of body size. As well as the evolution of new, large species through the Cenozoic Era, new small horses evolved[1]. Indeed up until about 10 million years ago there were species of horses that would have still weighed only about 75 kg, yet others weighed in at 400 kg. Such size increase that did occur would have happened either by an acceleration in the rate of growth or an extension of the period of growth – or both. In other words, a delay in the onset of maturity allows the period

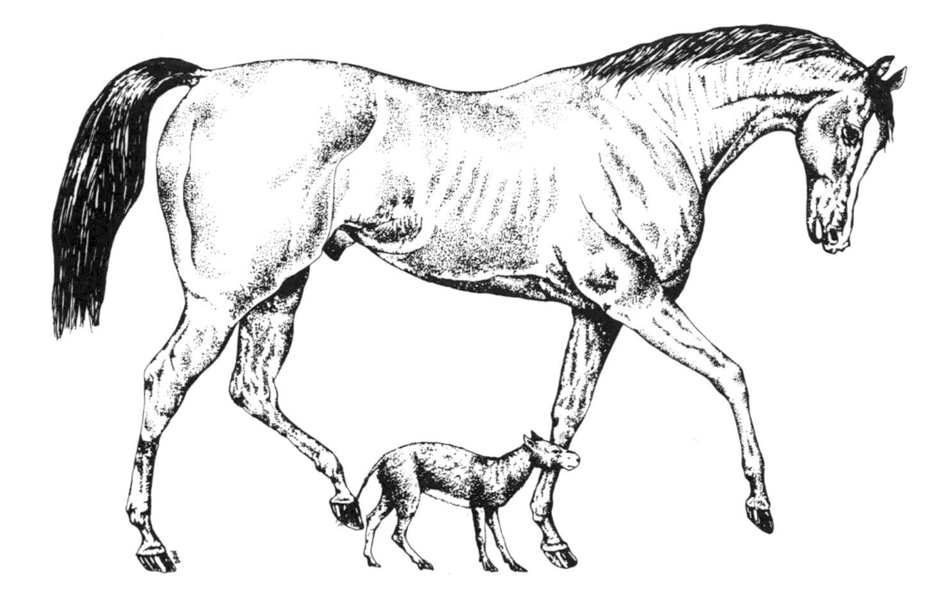

A *horse is a horse, of course of course.* Or *is it?* Hyracotherium, *the first primitive horse about the size of a fox terrier, scampers beneath a modern racehorse, that well-known triple* Derby *winner, 'Stillrunning'* (*genus* Equus).

of rapid, juvenile growth to carry on for longer. Direct evidence exists from the fossil record to indicate that Miocene horses, for example, that lived about 20 million years ago, reached sexual maturity at an earlier age than later horses. Extending the period of growth in later horses produced not only a larger body size, but also changes in shape of many parts of the body.

As organisms grow, different parts grow at different rates with respect to each other, and may themselves vary their growth rates during development. This results in distinct changes in shape during development. Dogs are a classic example of this. Compare the newborn puppy of a German Shepherd with the adult, and there are some pretty substantial differences. The puppy has a much larger head, relative to the rest of the body. Its limbs, on the other hand, are proportionately smaller. Its face is flatter, and during development the muzzle increases greatly in size and changes substantially in shape. This pattern holds, to varying degrees, for many mammals. As a result, changing the rates or periods of growth can induce substantial morphological changes.

Often, however, selection will target particular traits. We have seen in the previous chapter how selection in humans seems to have targeted the larger brain, larger body size, longer legs and longer developmental period. With this came developmental trade-offs – reduced jaw and teeth size, smaller gut, and so on. So it was with horses. The classic evolutionary change in horses during the Cenozoic was the evolution of the hoof at the expense of the other toes. Thus through a range of species there was an overall trend for a reduction in the number of toes in the fore and hindlimbs. This occurred by a reduction in size, brought on by a reduction of the growth rate of each of the toes, apart from one. This, on the other hand, underwent an increase in the amount of its growth, either an acceleration, or an increase brought about simply by the overall increase in body size, dragging this single toe along with it. A similar event happened in the evolution of extinct sthenurine kangaroos. They likewise ended up with a solitary toe – racehorses of the kangaroo world. South American liptoterns, that evolved some 50 million years ago, were also rather horse-like in appearance, and they too independently underwent a reduction in toe number from five to three to one in forms living 20 million years ago. In such cases we have more examples of classic developmental trade-offs – reduction in the number of toes (paedomorphosis), but a great increase (peramorphosis) in the remaining one to produce a hoof – essentially just an enormous toe nail.

Another aspect of human evolution, apart from critical developmental trade-offs, was the significance of the stretching out of successive developmental periods. It is quite possible that this same process is also responsible for perhaps the most classic example of paedomorphosis seen amongst living animals – the Mexican axolotl, *Ambystoma mexicanum*. Long touted as a classic example of 'neoteny', which is a reduction in the rate of growth, the appearance of the axolotl, looking for all the world like a juvenile, yet being sexually mature, does not arise from a change to its rate of growth. Whereas many salamanders undergo a metamorphosis from an aquatic juvenile form, which has gills, to a land-based, air-breathing preadult form, which then matures on land to become a land-based adult, the axolotl remains aquatic throughout its life. Consequently, as a sexually mature adult it retains the premetamorphic juvenile head, complete with external gill anatomy. To understand how this comes about we have to look not at how fast it grows, but the timing of transformation, or lack of transformation, from one developmental stage to another.

The ancestral developmental sequence, such as occurs in a related species, *Ambystoma tigrinum*, passes through embryonic, larval, metamorphic and postmetamorphic periods. Following metamorphosis the salamander leaves the water to live on land. The axolotl follows the same ancestral embryonic and larval developmental pattern to a point when the head undergoes no further shape change. It fails to undergo metamorphosis, which the ancestral form does when about six months old, but remains, Peter Pan-like, in a state of perpetual youth. There is no difference in time at which sexual maturity is initiated in either species, both occurring at about ten months[2]. What is happening is that growth and development of the body is quite dissociated from gonadal development, which determines when the species becomes sexually mature. By failure of the axolotl to metamorphose, the ancestral, premetamorphic larval features are retained by the descendent adult, so it remains living in an aquatic environment. Thus, like humans, the evolution of this phenomenon in the axolotl comes about by a delay in the transition from one growth stage to the next. In the case of the axolotl this transformation is delayed terminally, so that it just doesn't happen at all.

This same phenomenon can actually occur within some species of salamanders. Stephen Reilly of Ohio University found that in the living mole salamander *Ambystoma talpoideum* metamorphosis normally occurs during the first year, prior to the onset of sexual maturity[3]. However, in some populations there is a delay in the onset of metamorphosis for one or more years. However, the timing of onset of sexual maturity does not vary. As a result, as adults they retain some juvenile features, but eventually they do metamorphose.

Such changes to the time of transition from one growth stage to another are, I believe, an unheralded, but very important, aspect of evolution. For instance, a recent reinterpretation of yet another evolutionary 'classic', the extinct North American brontotheres, can be explained in much the same way. Brontotheres are an extinct group of perissodactyl mammals that lived during the Early Cenozoic. They are characterised by the attainment of huge body sizes for the time, and the presence in later, larger species, of bony horns that projected from the front of the skull. They were the subject of a classic monographic study published by Henry Osborn in 1929[4].

The earliest brontotheres that lived during the Early Eocene, some 50 million years ago, were small and lacked horns. Later species were larger and evolved horns that were disproportionately longer relative to increasing skull size. The largest of the brontotheres, species of *Brontotherium* that lived during the Oligocene Period, sported huge horns. These evolutionary trends in body and horn size became one of the classic examples of a vertebrate macroevolutionary trend, in particular

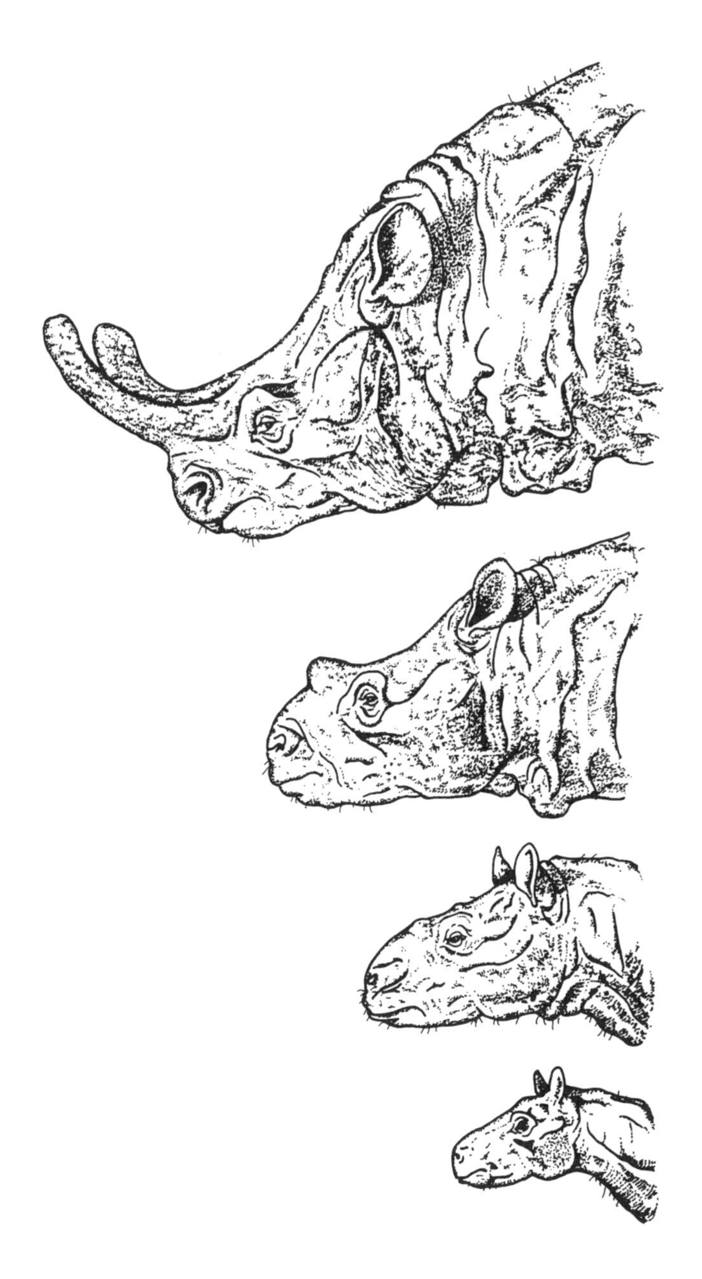

Evolution through taking ugly pills? No, not really, just that beauty is always in the eye of the beholder, especially if you were a brontothere. Progressive large size with increasing horn development was seen as a natural trend in the evolution of several groups of mammals arising from the stretching out of periods of growth and development.

one that evoked 'Cope's Rule', i.e. progressive increase in size. The evolution of large body and horn size were seen as being adaptive selection for massive size and structures that would have been used in head-to-head ramming during intraspecific combat.

In terms of the disproportionate changes in shape of the horns, these have long been cited as the classic example of allometric scaling, whereby a structure becomes larger relative to some other structure, in this case the horn relative to the skull. However, in 1985, Mike McKinney of the University of Tennessee and Robert Schoch of Yale University published a study that they had undertaken of the role of heterochrony in the evolution of the brontothere horns[5]. Using the detailed dimensions that Osborn had published they argued that the situation was far more complicated than just a simple case of allometric scaling. They calculated that for the horns to increase in size to the extent that they did, there was not only an extension of the period of growth, but that the growth rate was also accelerated, as well as horn growth being initiated progressively earlier in successive species. The result was that in successively younger species horns started growing earlier, grew faster and for a longer period. This, argued McKinney and Schoch, explained the brontotheres' enormous size.

It is 11 years after McKinney and Schoch's paper is published, and Osborn's data comes into its own again. This time the person interpreting the data is Gerald Bales of Western University of Health Sciences in Pomona, California. Bales has argued that an aspect of the way these horns may have developed that has not been considered in the past was the effect of changes to the life history stages, in particular changes to their duration[6]. Bales suggested that by comparison with a wide range of large modern mammals it is possible to estimate the times of gestation, and thus durations, of the embryonic and juvenile periods of growth. It is well known that a number of basic generalisations can be made concerning life history stages in living mammals. For instance, the adult size reflects the spacing of developmental events, including birth and onset of sexual maturity. Moreover, developmental phases occupy a constant proportion of an animal's lifespan. Thus larger species that live longer have longer gestation periods and take longer to attain sexual maturity than smaller forms. Another well-established constant is that the duration of embryonic development occupies about 2% of a mammal's lifespan, while sexual maturity is attained after approximately one-tenth of the animal's lifespan has been passed through. Bales therefore argued that using these values it could be argued that the different-sized brontotheres had life history stages of different durations. If this is beginning to sound familiar it should. Bales further suggests that horn growth may have accelerated late in development. Consequently, by extending the growth period this late acceleration could have proceeded for a longer period, producing larger horns.

Today many of us feel comfortable with the premise that fossils represent the remains of organisms that lived sometime in the past. We collect them because of their intrinsic interest and for what they can tell us about past life on this planet. Humans have been collecting fossils for a very long time. Amongst some of the earliest collected fossils is a fossil sea-urchin preserved in the middle of an Acheulian flint hand-axe found at Swanscombe in England[7]. The urchin nestles in the centre of the axe, having been carefully avoided by the manufacturer. What is remarkable is that this tool was manufactured about 200 000 years ago. Many fossil urchins have been recovered from archaeological sites.

In 1887 archaeologist Worthington G. Smith excavated a Bronze Age grave near Dunstable in England, and discovered the skeleton of a young woman and a baby. With them were more than 200 fossil sea-urchins, of the heart-shaped *Micraster* and the dome-shaped *Echinocorys*[8]. Long before their true nature as the remains of ancient forms of life was recognised, many kinds of fossils were collected and put to a variety of uses including, quite often, as grave goods. One of the main properties which fossils, like the sea-urchins, were thought to possess, was the power to confer luck or protection on their owners. This was often a two-edged sword. Pliny records how in Ancient Rome the Emperor Claudius was once confronted by an individual pleading some particular case. When quizzed as to what his arguments were, the hopeful soul produced a fossil urchin, arguing that the good fortune emanating from the object should be reason enough for him to win his case. Claudius, by all accounts, was not very impressed, having the litigant put to death for being so impertinent.

This belief in the good luck of fossils derives from much older beliefs, that these objects were somehow sent from the gods. One of their common names 'thunderstones' derives from the belief that they arrived on Earth from thunderstorms. Old flint tools were likewise considered to have been tossed to Earth by Thor during one his more cranky moods. At one particular early Iron Age (about 2500 years old) cremation burial site in Tunbridge Wells in Kent, England, a single flint *Micraster* and part of a worn Neolithic flint tool were found, both of which had been placed inside a pottery bowl.

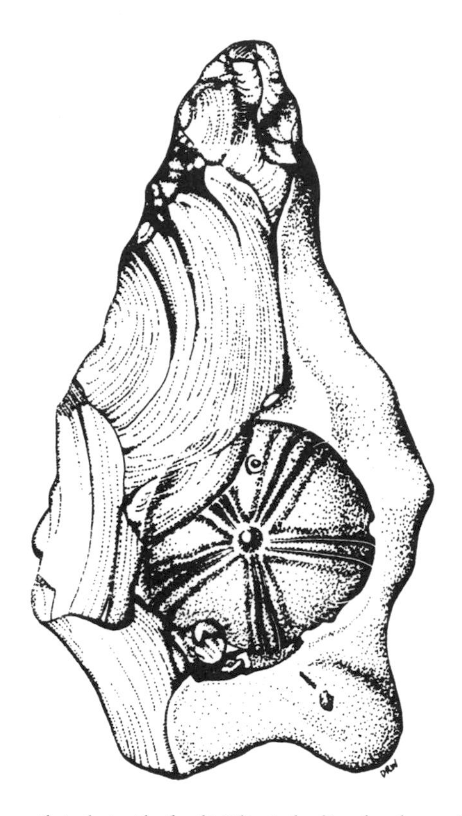

A fossil in the hand is worth tools in the bush! This Acheulian hand-axe, dated at around 200 000 years old, shows that even early hominids appreciated a good fossil. In as much as the fossil sea-urchins held great symbolism to these people, today such fossils symbolise the continuity and universality of evolution in our world around us. Today, fossils are more than just mere guides to knowledge of the past, they are vital tools for divining insights to predict future biotic patterns.

In the 1920s archaeological excavations at Whitehawk, in Brighton, on England's south coast yielded skeletons from two shallow graves. One grave had held the remains of a young woman, as did the other, along with an infant. Found with the skeletons were a few objects with which they had been buried, and which, presumably, must have meant a great deal to them during their short lives. Accompanying the skeletons of the woman and infant were two simple pieces of chalk, into which holes had been bored, apparently to allow her to thread twine through them to wear as a necklace, and the lower half of the radius of an ox. There was also one other prized possession – a fossil sea-urchin. The other grave with its lone skeleton contained just one object – also a fossil sea-urchin.

So, how did these women, thought to have been aged between 25 and 30, and the child, barely a few weeks old, meet their deaths? And of what significance to them, and to the woman and child buried at Dunstable, could fossil sea-urchins be in the afterlife? One suggestion that has been proposed is that the burials may have been part of a ritual. It has been argued that cannibalism and the eating of human brains took place at the Whitehawk encampment and was accompanied by ritualistic, perhaps even sacrificial, burial of the young women and the child: ritually killed and placed into a rough cist of chalk blocks.

Certainly, the Whitehawk encampment was not primarily a burial site, but an important meeting site. Why were just these three bodies buried here, unless they did indeed play their own, sacrificial part in a ritual. While it is possible that the emplacement of fossils with the dead bodies could have occurred because the objects themselves were of particular significance to the dead people, it would be unlikely for such an altruistic attitude to have been carried out by those who killed the women and child, if these people were sacrificially killed. If such were the case, a particularly potent symbolism may have been attached to the fossils. If they had belonged to the women, they may well also have had some symbolic, or 'religious' significance. Alternatively, perhaps they were mere curios, but ones to which they were particularly attached in life – evidence that humankind's propensity for collecting strange and interesting objects has a rich and prolonged history. Was the collecting instinct as much a part of the human's psyche 6000 years ago as it is today? Or was a much deeper significance attached to fossil sea-urchins by these early peoples? So is it a mere coincidence that the fossil urchins were buried with the women? These are just a few of the many cases that have been documented in the archaeological literature from all over Europe and the Near East for over 100 years of humans' fascination with fossil sea-urchins.

The idea that fossil urchins were 'thunderstones' that had fallen from the heavens had been a popular view in Roman times, and no doubt earlier. Pliny, writing in his *Natural History*, described how they were called 'brontia' and 'ombria', meaning 'thunder' and 'rainstorm', respectively, in Greek, and that they fell from the skies:

> Brontia is shaped in the manner of a Tortoise head: it falleth with a crack of thunder (as it is thought) from heaven; and if we will believe it, quencheth the fire of lightning . . . Ombria, which some call Notia, is said to fall from heaven in storm, showers of rain, and

> lightning, after the manner of other stones, called thereupon Ceraunia [the Greek for thunderbolt] and Brontia: and the like effects attributed to it.

Pliny described another legend involving sea-urchins. He recounts the tale of the Druid's 'snake egg'. According to this legend, many snakes twining together at midsummer make a ball out of froth, which they extrude. This was called *ovum anguinum* (literally, snake egg). The legend has it that if the object was able to be stolen from the snakes it acquired great magical properties. In particular it was said to ensure great success in battle or disputes[9].

Anselm de Boodt who, in addition to being a physician to Emperor Rudolph II of Prague, was a philosopher, theologian and naturalist, described two sorts of sea-urchins in his *Gemmarum et Lapidum historia*, published in 1609. His '*ova anguinum*', as he called them, he considered to be prize antidotes against poison. It is interesting to note that one of the most common species of *Micraster* was named *Micraster coranguinum* by Leske in 1778. Fossil sea-urchins, in particular *Micraster*, were widely used in folk medicine. In England John Woodward in his *Attempts Towards a Natural History of the Fossils of England*, which was published in 1729, recorded that urchins dug out of a chalk pit in Kent were highly prized as they contained very pure chalk within them. Known to him as 'chalk eggs', this fine chalk was said to be an excellent remedy for 'subduing acid humours of the stomach'[9].

The transformation of the significance of fossils to early cultures has been described by the late Kenneth Oakley as falling into three phases: the first, one of an attraction to the aesthetic appeal, or symbolic virtue, or some desired attribute that conveyed luck[10]. This was followed by a second, more advanced phase, when the fossils took on magic or mystic powers; and a third phase when their significance declined, as they passed into folklore as objects of good luck. Thus in nineteenth century southern England, and part of northern Europe, these shepherds' crowns, shepherds' knees, beggars' kneecaps, policemen's helmets, bishops' mitres or fairyloaves, as fossil sea-urchins were variously known, were thought to always keep a house in bread, and to stop milk from turning sour.

So these multipurpose objects, that rank as the oldest known objects that were collected by humans for something other than utilitarian purposes, achieved yet another use at the twilight of the nineteenth century and through the twentieth century: they became a classic demonstration of the activity of evolution. The original work by Rowe in 1899, followed by that of Kermack in 1954 and Nichols in 1959, showed how, in their opinion, subtle changes in the form of the urchin's test (shell) were adaptations that allowed later species to burrow deeper in the sediment. But again, such adaptively driven explanations never looked at the intrinsic driving forces producing the morphological variation that enable new morphologies to evolve allowing new niches to be invaded. To appreciate how this happened in *Micraster* we must turn to the study of other heart urchins. For this unravelled not only the underlying evolutionary mechanism, but also allowed this 'classic' example of evolution in *Micraster* to be reassessed and reinterpreted.

What enabled this to occur was a study that I carried out with my colleague Graeme Philip of the University of Sydney on younger heart urchins from Australia[11]. As we began categorising the different species of a heart urchin called *Schizaster*, it soon became apparent to us that the earliest species, from 55 million-year-old rocks from northwest Australia, were very different from much younger species. Moreover, the early species were urchins that had burrowed in quite coarse limy sediment, whereas species living from about 20 million years ago to the present day had evolved the ability to live in very fine-grained sediments. To do this urchins need quite sophisticated anatomical features that allow them to construct a funnel from their deep burrow, up to the sediment surface. Down this funnel flows life-giving oxygen and food-bearing water.

In a coarse sediment it is not too much of a problem, because water can percolate between sediment grains. In a mud, however, it is another matter. A clear funnel must be constructed and maintained. Examining the differences between the earliest and latest species, and looking at how they changed through a series of intermediate species that lived in Late Eocene (40 million years ago) to Late Oligocene (25 million years ago) times, allowed us to unravel what adaptations had evolved, and how. In the earliest species the heart shape was not pronounced. The five grooves that radiate out from the centre of the urchin's test were shallow. The pores, through which passed tiny tube feet used for respiration and for secreting mucus to line the funnel, were relatively few in number. These were all trademarks of species adapted to living in shallow water, high energy, coarse sediment environments.

When the development of individual specimens was studied it could be seen that such changes that did occur from early juvenile to adult were not profound. The grooves (or ambulacra) became a little

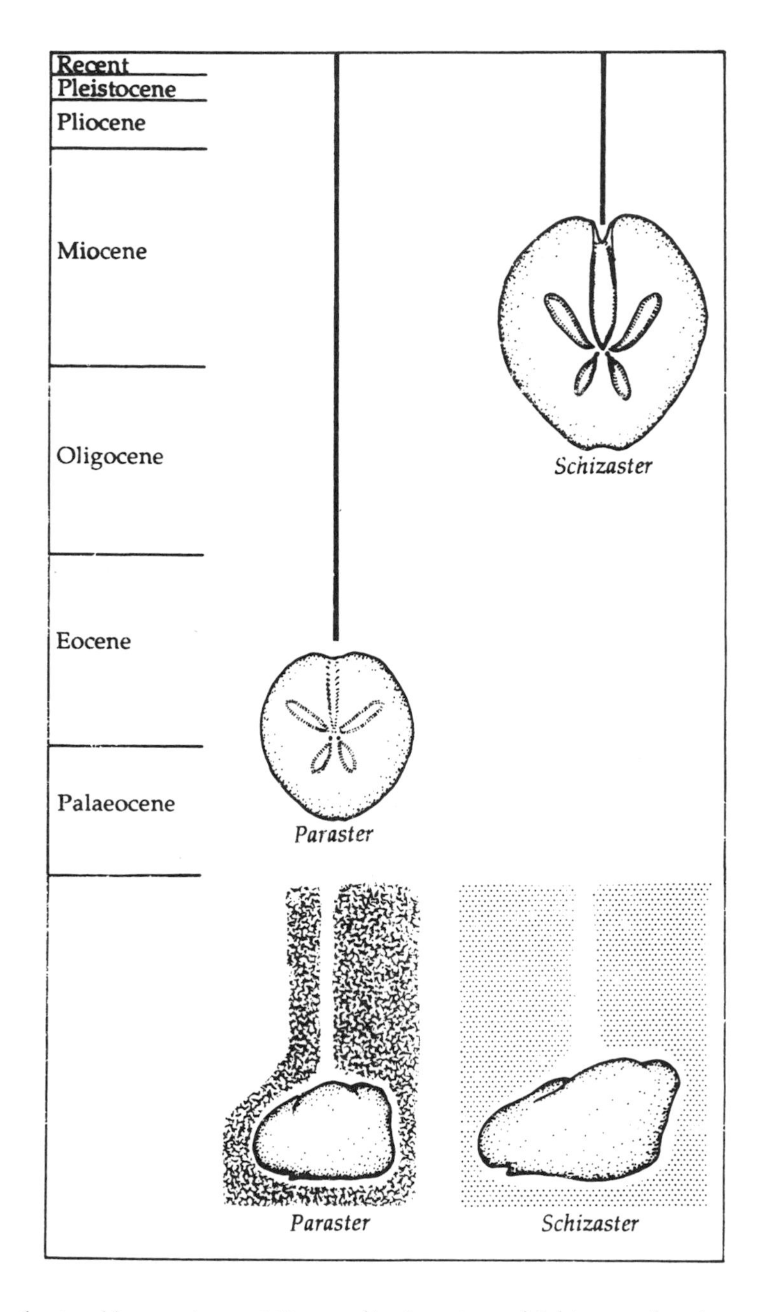

Bury my heart and leave me in peace! Heart *urchins* Paraster *and* Schizaster *show shapes reflecting their adaptations towards living in sediments of different grain size. As with many lineages of heart urchins, evolutionary trends were from inhabiting coarse-grained sediments in shallow water to burrowing deep in fine-grained silts and muds in deeper water.*

deeper, but other changes in shapes and proportions were minimal. At the other extreme, however, were living species, like *Schizaster myorensis* that Graeme and I described from specimens living in mud off the Queensland coast. The adults of the heart urchins had wedge-shaped tests, and ambulacra that were deeply sunken. The heart shape of the test was pronounced, with a deep groove at the front. The whole morphology was adapted to utilising a very focused source of life-giving water as efficiently as possible.

Some tube feet were used to secrete mucus to line the funnel to the surface. In *Schizaster myorensis* these were crowded into the test. To resolve how these changes occurred we looked at juveniles. Although paper thin, enough had been collected for us to realise that the juveniles of these later, mud-dwelling species were just like the juveniles, and to a lesser extent the adults, of our 55 million-year-old *Schizaster carinatus*. Evolution in this lineage had therefore been driven, like many dinosaurs, by peramorphosis – an acceleration in growth rates and/or an extension in the period of growth. Thus pre-existing shape changes and extent of production of tube feet were accentuated, producing novel morphologies that became established in new, in this case muddier, niches. Species intermediate in age were intermediate in morphology, and living in sediments of intermediate grain size. So here we could see an evolutionary trend developing by extending the ancestral developmental history and evolving along an environmental gradient of coarse to fine-grained sediments. Such evolutionary trends in species becoming more and more peramorphic, or more and more paedomorphic, I was later to call 'peramorphoclines' and 'paedomorphoclines'[12]. Many other examples of these, in a wide range or organisms from trilobites to fish to starfish to trees, have been described by scientists over the last decade.

Picking up on the patterns of evolution portrayed by *Schizaster*, Andrew Smith of the Natural History Museum in London applied it to the 'classic' evolutionary story of *Micraster* and found it could be interpreted in another way[13]. For rather than the changes being adaptations to burrowing deeper in the sediment, they, like *Schizaster*, were related to the sediment grain size. The changes seen in *Schizaster* had been presaged tens of millions of years earlier by species of *Micraster*. And in all likelihood the same internal driving forces of increased amount of change during development generated the new morphologies that were able to colonise finer-grained sediments.

Well, that is the story to date of the evolution of *Micraster* and of *Schizaster* and the countless other examples of evolution that the fossil record has revealed. At least, that is how we interpret it today. No doubt, as new discoveries are made, and new ideas about how evolution works evolve, how we interpret the changes that the fossil record has garnered for us over three and half billion years will be modified and extended. But then that is the wonder of palaeontology and the impetus that keeps us digging for new specimens and delving for new ideas. The research funds might be drying up and the mountains of paperwork might be about to suffocate us, but – who knows – the next time that we get out into the field, that rock we hit might just be the one that contains another critical piece of the great jigsaw of life.

19

Final Thoughts

'Greetings from the wilds of Patagonia – have been lucky this last year – two large carnivorous dino skeletons – and they are weirdly decorated. More about this soon. Pat and Tom.'

It is now February 1998, and yesterday we received this message on a postcard from Pat Vickers-Rich and Tom Rich. Expanding their horizons from southern Australia they have branched out into the rich Cretaceous dinosaur country of Patagonia in South America. More new finds and other pieces in the great jigsaw of the evolution of life. Since we submitted the manuscript for this book to the publishers a few months ago, many more new finds and new interpretations of ancient life have been made. It's a full time job just keeping track of it all.

It has just been reported that the oldest fossil ants have been discovered in amber from New Jersey. Dated at 92 million years old, this is pushing back the origin of ants deeper into the Cretaceous. The authors of the article suggest ants evolved about 130 million years ago[1]: yet another major group of organisms to appear in the Early Cretaceous. What is it about the Cretaceous? When it comes to new fossil discoveries it keeps cropping up: amazing new fossil birds from Spain, Madagascar and China; 'feathered' dinosaurs; new early plant and insect discoveries; new finds of early mammals. Cretaceous times saw the explosive radiation of so many groups: flowering plants; birds, modern insect groups, frogs and, to a lesser extent, mammals. In a recent article Robert Sloan of the University of Minnesota has argued that the Cretaceous was, in many ways, a very unusual time[2]. Unusual in that the world experienced a number of major physical changes as well as being a period of great evolutionary change.

The impact that the physical changes had on the evolution of life has, in all likelihood, been profound. Whereas throughout this book we have looked at various groups of animals and plants largely in isolation from each other, in the real world many of these groups were interacting on a day-to-day basis as parts of complex ecosystems. Consequently the evolutionary histories of many of these groups are inextricably intertwined as these ecosystems evolved. So how can the evolution of flowers, dinosaurs, birds, marine reptiles and fishes, as well as the first ants, be interrelated?

Between 145 and 66 million years ago the great southern supercontinent of Gondwana fell apart, as a result of a doubling of the pace of sea floor spreading. A consequence of this was a doubling of the rate of output of carbon dioxide into the atmosphere. Thus a six-fold increase in carbon dioxide in the atmosphere created a greenhouse world, and warming of the planet ensued. This extra carbon dioxide led to a great evolutionary radiation in plants, especially the flowering plants. With greater photosynthesis and burial of organic matter as oil and coal, oxygen levels in the atmosphere increased, to as much as 35%. As we have pointed out earlier, the outcome of the evolutionary radiation of flowering plants was the coevolution of a great diversity of birds and insects – major pollinators. The four largest groups of insects – the beetles, flies, wasps/bees/ants and moths/butterflies evolved to such an extent that they now make up 90% of all living insects[2]. With so much potential new food around, insect eaters, such as frogs, salamanders, lizards and early members of modern mammal groups, evolved rapidly.

Sloan considers that the evolutionary history of dinosaurs was specifically tied to increasing oxygen levels. He argues that the greater amount of oxygen in the atmosphere allowed the evolution of large body size in dinosaurs (as it probably did the huge pterosaurs that lived in the Cretaceous). Likewise, Sloan links major extinctions, including the terminal one at the end of the Cretaceous, to falls in oxygen levels. Dinosaurs were probably not very efficient breathers[3]. And this is where the pubic boot comes in (see chapter 14). In a recent article John Ruben and colleagues from Oregon State University have suggested that dinosaurs may have breathed like crocodiles, utilising a 'hepatic-piston diaphragm' method of breathing[4]. Basically diaphragmatic muscles inserted onto the pubic boot attached to the back of the liver. Contraction in a piston-like manner pulled air into the lungs. The argument goes, that the bigger the pubic boot, the more air could be sucked in, so potentially the larger the body size that could be supported. This would have been fine while oxygen levels were high, but when they dropped towards the end of the Cretaceous, large dinosaurs would basically have been unable suck in enough air. The more efficient birds and mammals, however, with their smaller body sizes, would have been able to cope with lower oxygen levels.

In the sea the higher carbon dioxide levels led to vast blooms of phytoplankton, resulting in vast chalk deposits being laid down in the late Cretaceous. This in turn resulted in increases in small crustaceans and other invertebrates that fed on the plankton, then to increases in those animals that fed on these invertebrates. The great evolutionary radiation in modern teleost fishes took place during this time, along with that of modern sharks and rays.

As our levels of knowledge of ancient ecosystems develop, as more fossil discoveries are made and interpreted, along with our understanding of changes to the physical world, the more we will be able to unravel the past patterns of life on this planet. Only then will we be able fully to understand and appreciate the impact that we, as relative newcomers to the Earth's three and half billion-year-old evolutionary heritage, will have on the evolution of ecosystems of the future.

References

Chapter 1

1. Groves, D.I., Dunlop, J.S.R. and Buick, R. 1981. An early habitat of life. *Scientific American* **29**: 183–206.
2. Buick, R., Thornett, J.R., McNaughton, N.J., Smith, J.B., Barley, M.E. and Savage, M. 1995. Records of emergent continental crust ~3.5 billion years ago in the Pilbara craton of Australia. *Nature* **375**: 574–577.
3. McNamara, K.J. and Awramik, S.M. 1994. Stromatolites: a key to understanding the early evolution of life. *Science Progress* **77**: 1–20.
4. Walcott, C.D. 1915. Discovery of Algonkian bacteria. *Proceedings of the National Academy of Sciences* **1**: 256–257.
5. Schopf, J.W. 1994. The early evolution of Life: solution to Darwin's dilemma. *Trends in Ecology and Evolution* **9**: 375–377.
6. Awramik, S.M., Schopf, J.W.E. and Walter, M.R. 1983. Filamentous fossil bacteria from the Archean of Western Australia. *Precambrian Research* **20**: 357–374.
7. Schopf, J.W. 1993. Microfossils of the early Archean Apex Chert: new evidence of the antiquity of life. *Science* **260**: 640–646.
8. Gould, S.J. 1996. *Life's Grandeur*. Random House, London.
9. Mojzsis, S.J., Arrhenius, G., McKeegan, K.D., Harrison, T.M., Nutman, A.P. and Friend, C.R.L. 1996. Evidence for life on Earth before 3,800 million years ago. *Nature* **384**: 55–59.
10. Cohen, P. 1996. Let there be life. *New Scientist* 6 July, 1996: 22–27.
11. Glover, J.E. 1992. Sediments of Early Archaean coastal plains: a possible environment for the origin of life. *Precambrian Research* **56**: 159–166.
12. Wächtershäuser, G. 1988. Before enzymes and templates: theory of surface metabolism. *Microbiological Review* **52**: 452–484.
13. Grey, K. and Thorne, A.M. 1985. Biostratigraphic significance of stromatolites in upward shallowing sequences in the Early Proterozoic Duck Creek Dolomite, Western Australia. *Precambrian Research* **29**: 183–206.
14. Wray, G.A., Levinton, J.S. and Shapiro, L.H. 1996. Molecular evidence

for deep Precambrian divergences among metazoan phyla. *Science* **274**: 568–572.

15. Ayala, F.J., Rzhetsky, A. and Ayala, F.J. 1998. Origin of the metazoan phyla: molecular clocks confirm paleontological estimates. *Proceedings of the National Academy of Sciences* **95**: 606–611.

Chapter 2

1. Han, T.-M. and Runnegar, B. 1992. Megascopic eukaryotic algae from the 2.1-billion-year-old Negaunee Iron-Formation. *Science* **257**: 232–235.
2. Grey, K. and Williams, I.R. 1990. Problematic bedding-plane markings from the Middle Proterozoic Manganese Subgroup, Bangemall Basin, Western Australia. *Precambrian Research* **46**: 307–327.
3. Fedonkin, M.A., Yochelson, E.L. and Horodyski, R.J. 1994. Ancient Metazoa. *National Geographic Research and Exploration* **10**: 200–223.
4. Runnegar, B. 1995. Vendobionta or Metazoa? Developments in the understanding of the Ediacaran 'fauna'. *Neues Jahrbuch für Geologie und Paläontologie Abhandlungen* **195**: 305–318.
5. Hoffman, H.J., Narbonne, G.M. and Aitken, J.D. 1990. Ediacaran remains from intertillite beds in northwestern Canada. *Geology* **18**: 1199–1202.
6. Fedonkin, M.A. and Runnegar, B.N. 1992. Proterozoic metazoan trace fossils. *In* J.W. Schopf and C. Klein (eds), *The Proterozoic Biosphere*, pp. 389–410, Cambridge University Press, Cambridge.
7. Crimes, T.P. 1994. The period of early evolutionary failure and the dawn of evolutionary success: the record of biotic changes across the Precambrian–Cambrian boundary. *In* S.K. Donovan (ed.), *The Palaeobiology of Trace Fossils*, pp. 105–133, John Wiley & Sons, Chichester.
8. Webby, B.D. 1970. Late Precambrian trace fossils from New South Wales. *Lethaia* **3**: 79–109.
9. Wray, G.A., Levinton, J.S. and Shapiro, L.H. 1996. Molecular evidence for deep Precambrian divergences among metazoan phyla. *Science* **274**: 568–572.
10. Davidson, E.H., Peterson, K.J. and Cameron, R.A. 1995. Origin of bilateral body plans: evolution of developmental regulatory mechanisms. *Science* **270**: 1319–1325.
11. Xiao, S., Zhang, Y. and Knoll, A.H. 1998. Three-dimensional preservation of algae and animal embryos in a Neoproterozoic phosphorite. *Nature* **391**: 553–558.
12. Ford, T.D. 1958. Precambrian fossils from Charnwood Forest. *Proceedings of the Yorkshire Geological Society* **31**: 211–217.

13. Glaessner, M.F. 1984. *The Dawn of Animal Life*. Cambridge University Press, Cambridge.
14. Pflug, H.D. 1973. Zur fauna der Nama-Schichten in Südwest-Afrika. IV. Mikroscopische anatomie der petal-organisme. *Palaeontolographica* **B144**: 166–202.
15. Seilacher, A. 1989. Vendozoa: organismic construction in the Proterozoic. *Lethaia* **22**: 229–239.
16. Seilacher, A. 1992. Vendobionta and Psammocorallia: lost constructions of Precambrian evolution. *Journal of the Geological Society, London* **149**: 607–613.
17. Fedonkin, M.A. 1985. Precambrian metazoans: the problems of preservation, systematics and evolution. *Philosophical Transactions of the Royal Society, London* **B311**: 27–45.
18. Conway Morris, S. 1993. Ediacaran-like fossils in Cambrian Burgess Shale-type faunas of North America. *Palaeontology* **36**: 593–635.
19. Fedonkin, M.A. and Waggoner, B.M. 1997. The Late Precambrian fossil *Kimberella* is a mollusc-like bilateran organism. *Nature* **388**: 868–871.
20. Retallack, G.J. 1994. Were the Ediacaran fossils lichens? *Paleobiology* **20**: 523–544.

Chapter 3

1. Zhang W.-T. and Hou X.-G. 1985. Preliminary notes on the occurrence of the unusual trilobite *Naraoia* in Asia. *Acta Palaeontologica Sinica* **24**: 591–595 [in Chinese with English summary].
2. Luo H.-L. and Zhang S.-S. 1986. Early Cambrian vermes and trace fossils from Jinning-Anning region, Yunnan. *Acta Palaeontologica Sinica* **25**: 307–311.
3. Conway Morris, S. 1998. *The Crucible of Creation: the Burgess Shale and the Rise of Animals*. Oxford University Press, Oxford.
4. Ramsköld, L. and Hou X.-G. 1991. New Early Cambrian animal and onychophoran affinities of enigmatic metazoans. *Nature* **351**: 225–228.
5. Hou X.-G., Ramsköld, L. and Bergström, J. 1991. Composition and preservation of the Chengjiang fauna – a Lower Cambrian soft-bodied biota. *Zoologica Scripta* **20**: 395–411.
6. Ramsköld, L. 1992. The second leg row of *Hallucigenia* discovered. *Lethaia* **25**: 221–224.
7. Budd, G. 1993. A Cambrian gilled lobopod from Greenland. *Nature* **364**: 709–711.
8. Whittington, H.B. and Briggs, D.E.G. 1985. The largest Cambrian animal *Anomalocaris*, Burgess Shale, British Columbia. *Philosophical Transactions of the Royal Society of London* **B309**: 569–609.

9. Whittington, H.B. 1975. The enigmatic animal *Opabinia regalis*, Middle Cambrian, Burgess Shale, British Columbia. *Philosophical Transactions of the Royal Society of London* **B271**: 1–43.
10. Conway Morris, S. 1993. The fossil record and the early evolution of the Metazoa. *Nature* **361**: 219–225.
11. Chen J.-Y., Zhou G.-Q. and Ramsköld, L. 1995. A new Early Cambrian onycophoran-like animal, *Paucipodia* gen. nov., from the Chengjiang fauna, China. *Transactions of the Royal Society of Edinburgh* **85**: 275–282.
12. Chen J.-Y., Dzik, J., Edgecombe, G.D., Ramsköld, L. and Zhou G.-Q. 1995. A possible early Cambrian chordate. *Nature* **377**: 720–722.
13. Conway Morris, S. 1989. The persistence of Burgess shale-type faunas: implications for the evolution of deeper-water faunas. *Transactions of the Royal Society of Edinburgh: Earth Sciences* **80**: 271–283.
14. Shu, D.-G., Conway Morris, S. and Zhang, X.-L. 1996. A *Pikaia*-like chordate from the Lower Cambrian of China. *Nature* **384**: 157–158.
15. Brasier, M., Green, O. and Shields, G. 1997. Ediacaran sponge spicule clusters from southwest Mongolia and the origins of the Cambrian fauna. *Geology* **25**: 303–306.
16. Gehling, J.G. and Rigby, J.K. 1996. Long expected sponges from the Neoproterozoic Ediacara Fauna of South Australia. *Journal of Paleontology* **70**: 185–195.
17. Hou X.-G. 1987. Two new arthropods from Lower Cambrian Chengjiang, eastern Yunnan. *Acta Palaeontologica Sinica* **26**: 236–256 [in Chinese with English summary].
18. Chen J.-Y., Edgecombe, G.D. and Ramsköld, L. 1995. Head segmentation in Early Cambrian *Fuxianhuia*: implications for arthropod evolution. *Science* **268**: 1339–1343.
19. Conway Morris, S. 1979. The Burgess Shale (Middle Cambrian) fauna. *Annual Reviews of Ecology and Systematics* **10**: 327–349.
20. Babcock, L.E. 1993. Trilobite malformations and the fossil record of behavioral asymmetry. *Journal of Paleontology* **67**: 217–229.

Chapter 4

1. Long, J.A. 1995. *The Rise of Fishes. 500 million years of Evolution.* University of New South Wales Press, Sydney; Johns Hopkins University Press, Baltimore.
2. Shu, D.-G., Conway Morris, S. and Zhang, X.-L. 1996. A *Pikaia*-like chordate from the Lower Cambrian of China. *Nature* **384**: 157–158.
3. Long, J.A. and Burrett, C.F. 1989. Tubular phosphatic microproblematica from the Early Ordovician of China. *Lethaia* **22**: 439–446.
4. Sansom, I.J., Smith, M.P., Armstrong, H.A. and Smith, M.M. 1992.

Presence of the earliest vertebrate hard tissues in conodonts. *Science* **256**: 1308–1311.

5. Purnell, M. 1995. Microwear on conodont elements and macrophagy in the first vertebrates. *Nature* **374**: 798–800.
6. Briggs, D.E.G., Clarkson, E.N.K. and Aldridge, R.J. 1983. The conodont animal. *Lethaia* **16**: 1–14.
7. Gabbott, S.E., Aldridge, R.J. and Theron, J.N. 1995. A giant conodont with preserved muscle tissue from the Upper Ordovician of South Africa. *Nature* **374**: 800–803.
8. Janvier, P. 1995. Conodonts join the club. *Nature* **374**: 761–762.
9. Ritchie, A. and Gilbert-Tomlinson, J. 1977. First Ordovician vertebrates from the southern hemisphere. *Alcheringa* **1**: 351–368.
10. Gagnier, P.Y. 1995. Ordovician vertebrates and agnathan phylogeny. *Bulletin du Muséum national d'Histoire naturelle* **17**: 1–37.
11. Young, G.C. 1997. Ordovician microvertebrate remains from the Amadeus Basin, Central Australia. *Journal of Vertebrate Paleontology* **17**: 1–25.
12. Young, G.C., Karatajute-Talimaa, V.N. and Smith, M.M. 1996. A possible Late Cambrian vertebrate from Australia. *Nature* **383**: 810–812.

Chapter 5

1. Trewin, N.H. and McNamara, K.J. 1995. Arthropods invade the land: trace fossils and palaeoenvironments of the Tumblagooda Sandstone (?late Silurian) of Kalbarri, Western Australia. *Transactions of the Royal Society of Edinburgh* **85**: 177–210.
2. Johnson, E.W., Briggs, D.E.G., Suthren, R.J., Wright, J.L. and Tunnicliff, S.P. 1994. Non-marine arthropod traces from the subaerial Ordovician Borrowdale Volcanic Group, English Lake District. *Geological Magazine* **131**: 395–406.
3. Gray, J. 1985. The microfossil record of early land plants: advances in understanding of early terrestrialization, 1970–1984. *Philosophical Transactions of the Royal Society of London* **B309**: 167–195.
4. Retallack, G.J. and Feakes, C.R. 1987. Trace fossil evidence for Late Ordovician animals on land. *Science* **235**: 61–63.
5. Sherwood-Pike, M.A. and Gray, J. 1985. Silurian fungal remains: oldest records of the class Ascomycetes. *Lethaia* **18**: 1–20.
6. Trewin, N.H. 1996. Preface to 'On Old Red Sandstone plants showing structure, from the Rhynie Chert Bed, Aberdeenshire', by R. Kidston and W.H. Lang. *Transactions of the Royal Society of Edinburgh* **87**: 423–425.
7. Jeram, A.J., Selden, P.A. and Edwards, D. 1990. Land animals in the

Silurian: arachnids and myriapods from Shropshire, England. *Science* **250**: 658–661.
8. Dunlop, J.A. 1996. A trigonorabid arachnid from the Upper Silurian of Shropshire. *Palaeontology* **39**: 605–614.
9. Shear, W.A., Bonamo, P.M., Grierson, J.D., Rolfe, W.D.I., Smith, E.L. and Norton, R.A. 1984. Early land animals in North America: evidence from Devonian age arthropods from Gilboa. *Science* **224**: 492–494.
10. Selden, P.A., Shear, W.A. and Bonamo, P.M. 1991. A spider and other arachnids from the Devonian of New York, and reinterpretations of Devonian Araneae. *Palaeontology* **34**: 241–281.
11. Shear, W.A. 1991. The early development of terrestrial ecosystems. *Nature* **351**: 283–289.

Chapter 6

1. Long, J.A. 1995. *The Rise of Fishes. 500 million years of Evolution.* University of New South Wales Press, Sydney; Johns Hopkins University Press, Baltimore.
2. Ritchie, A. 1985. *Arandaspis prionotolepis*. The Southern four-eyed fish. *In* P. Rich and G. van Tets (eds), *Kadimakara*, pp. 95–106, Pioneer Design Studios, Lilydale.
3. Gagnier, P.-Y. 1989. The oldest vertebrate: a 470 million year old jawless fish, *Sacabambaspis janvieri*, from the Ordovician of Bolivia. *National Geographic Research* **5**: 250–253.
4. Elliott, D.K. 1987. A reassessment of *Astraspis desiderata*, the oldest North American vertebrate. *Science* **237**: 190–192.
5. Forey, P.L. and Janvier, P. 1993. Agnathans and the origin of jawed vertebrates. *Nature* **361**: 129–134.
6. Van der Brughen, W. and Janvier, P. 1993. Denticles in thelodonts. *Nature* **364**: 107.
7. Wilson, M.V.H. and Caldwell, M.W. 1993. New Silurian and Devonian fork-tailed 'thelodonts' are jawless vertebrates with stomachs and deep bodies. *Nature* **361**: 442–444.
8. Long, J.A. and Young, G.C. 1995. New sharks from the Middle–Late Devonian Aztec Siltstone, southern Victoria Land, Antarctica. *Records of the Western Australian Museum* **18**: 287–308.
9. Forey, P.L. 1980. *Latimeria*: a paradoxical fish. *Proceedings of the Royal Society of London* **B208**: 369–384.
10. Young, G.C. 1986. The relationships of placoderm fishes. *Zoological Journal of the Linnean Society* **88**: 1–57.
11. Long, J.A. 1995. A new plourdosteid arthrodire from the Late Devonian

Gogo Formation, Western Australia: systematics and phylogenetic implications. *Palaeontology* **38**: 1–24.
12. Denison, R.H. 1941. The soft anatomy of *Bothriolepis*. *Journal of Paleontology* **15**: 553–561.
13. Long, J.A. 1993. Cranial ribs in Devonian lungfishes and the origin of dipnoan air-breathing. *Memoirs of the Association of Australasian Palaeontologists* **15**: 199–210.
14. Warren, A., Jupp, R. and Bolton, B. 1986. Earliest tetrapod trackway. *Alcheringa* **10**: 183–186.
15. Long, J.A. 1993. Early–Middle Palaeozoic vertebrate extinction events. *In* J.A. Long (ed.), *Palaeozoic Vertebrate Biostratigraphy and Biogeography*, pp. 54–66, Belhaven Press, London.

Chapter 7

1. McNamara, K.J. and Trewin, N.H. 1993. A euthycarcinoid arthropod from the Silurian of Western Australia. *Palaeontology* **36**: 319–335.
2. Schram, F.R. 1971. A strange arthropod from the Mazon Creek of Illinois and the trans Permo-Triassic Merostomoidea (Trilobitoidea). *Fieldiana: Geology* **20**: 85–102.
3. Bergström, J. 1980. Morphology and systematics of early arthropods. *Abhandlungen des Naturwissenschaftlichen Vereins in Hamburg* **23**: 7–42.
4. Schram, F.R. and Rolfe, W.D.I. 1982. New euthycarcinoid arthropods from the Upper Pennsylvanian of France and Illinois. *Journal of Paleontology* **56**: 1434–1450.
5. Freyer, G. 1996. Reflections on arthropod evolution. *Biological Journal of the Linnean Society* **56**: 1–55.
6. McNamara, K.J. 1997. *Shapes of Time*. Johns Hopkins University Press, Baltimore.
7. Robison, R.A. 1990. Earliest-known uniramous arthropod. *Nature* **343**: 163–164.
8. Whalley, P. and Jarzembowski, E.A. 1981. A new assessment of *Rhyniella*, the earliest known insect, from the Devonian of Rhynie, Scotland. *Nature* **291**: 317.
9. Averof, M. and Cohen, S.M. 1997. Evolutionary origin of insect wings from ancestral gills. *Nature* **385**: 627–630.
10. Marden, J.H. and Kramer, M.G. 1995. Locomotor performance of insects with rudimentary wings. *Nature* **377**: 332–334.
11. Carroll, S.B., Weatherbee, S.D. and Langeland, J.A. 1995. Homeotic genes and the regulation and evolution of insect wing number. *Nature* **375**: 58–61.

Chapter 8

1. Long, J.A. 1995. *The Rise of Fishes. 500 million years of Evolution*. University of New South Wales Press, Sydney; Johns Hopkins University Press, Baltimore.
2. Long, J.A. 1985. A new osteolepidid fish from the Upper Devonian Gogo Formation, Western Australia. *Records of the Western Australian Museum* **12**: 361–377.
3. Rosen, D.E., Forey, P.L., Gardiner, B.G. and Patterson, C. 1981. Lungfishes, tetrapods, palaeontology and plesiomorphy. *Bulletin of the American Museum of Natural History* **167**: 159–276.
4. Jarvik, E. 1980. *Basic Structure and Evolution of Vertebrates*. Vols 1, 2. Academic Press, London and New York.
5. Long, J.A. 1988. Late Devonian fishes from the Gogo Formation, Western Australia. *National Geographic Research* **4**: 436–450.
6. Long, J.A., Campbell, K.S.W. and Barwick, R.E. 1997. Osteology and functional morphology of the osteolepiform fish *Gogonasus andrewsae* Long, 1985, from the Upper Devonian Gogo Formation, Western Australia. *Records of the Western Australian Museum, Supplement* **53**: 1–90.
7. Lebedev, O.A. 1990. Tulerpeton, l'animal à six doigts. *La Recherche* **225**: 1274–1275.
8. Campbell, K.S.W. and Bell, M.W. 1977. A primitive amphibian from the Late Devonian of New South Wales. *Alcheringa* **1**: 369–381.
9. Schultze, H.-P. and Arsenault, M. 1985. The panderichthyid fish *Elpistostege*: a close relative of tetrapods? *Palaeontology* **28**: 293–310.
10. Clack, J.A. 1988. New material of the early tetrapod *Acanthostega* from the Upper Devonian of East Greenland. *Palaeontology* **31**: 699–724.
11. Clack, J.A. 1989. Discovery of the earliest tetrapod stapes. *Nature* **342**: 425–430.
12. Coates, M.I. and Clack, J.A. 1990. Polydactyly in the earliest known tetrapod limbs. *Nature* **347**: 66–69.
13. Coates, M.I. and Clack, J.A. 1991. Fish-like gills and breathing in the earliest known tetrapod. *Nature* **352**: 234–236.
14. Ahlberg, P.E. 1991. Tetrapod or near tetrapod fossils from the Upper Devonian of Scotland. *Nature* **354**: 298–301.
15. Ahlberg, P.E., Luksevics, E. and Lebedev, O. 1994. The first tetrapod finds from the Devonian (Upper Famennian) of Latvia. *Philosophical Transactions of the Royal Society of London* **B343**: 303–328.
16. Ahlberg, P.E. 1995. *Elginerpeton pancheni* and the earliest tetrapod clade. *Nature* **373**: 420–425.
17. Daeschler, E.B., Shubin, N.H., Thomson, K.S. and Amaral, W.W. 1994. A Devonian tetrapod from North America. *Science* **265**: 639–642.

18. Ahlberg, P.E., Clack, J.A. and Luksevics, E. 1996. Rapid braincase evolution between *Panderichthys* and the earliest tetrapods. *Nature* **381**: 61–64.
19. Young, G.C., Long, J.A. and Ritchie, A. 1992. Crossopterygian fishes from the Devonian of Antarctica: Systematics, relationships and biogeographic significance. *Records of the Australian Museum, Supplement* **14**: 1–77.
20. Schultze, H.-P. 1984. Juvenile specimens of *Eusthenopteron foordi* Whiteaves 1881 (osteolepiform rhipidistian, Pisces), from the Upper Devonian of Miguashua, Quebec, Canada. *Journal of Vertebrate Paleontology* **4**: 1–16.
21. Long, J.A. 1990. Heterochrony and the origin of tetrapods. *Lethaia* **23**: 157–166.
22. McKinney, M.L. and McNamara, K.J. 1991. *Heterochrony: the Evolution of Ontogeny*. Plenum Press, New York.
23. Jarvik, E. 1996. The Devonian tetrapod *Ichthyostega*. *Fossils and Strata* **40**: 1–206.
24. Coates, M.I. 1996. The Devonian tetrapod *Acanthostega gunnari* Jarvik: postcranial anatomy, basal tetrapod interrelationships and patterns of skeletal development. *Transactions of the Royal Society of Edinburgh, Earth Sciences* **87**: 363–421.
25. Shubin, N., Tabin, C. and Carroll, S. 1997. Fossils, genes and the evolution of animal limbs. *Nature* **388**: 639–648.
26. Sordino, P., van der Hoeven, F. and Duboule, D. 1995. Hox gene expression in teleost fins and the origin of vertebrate digits. *Nature* **375**: 678–681.
27. Ahlberg, P. and Milner, A.R. 1994. The origin and early diversification of tetrapods. *Nature* **368**: 507–512.
28. Thulborn, T., Warren, A., Turner, S. and Hamley, T. 1996. Early Carboniferous tetrapods in Australia. *Nature* **381**: 777–780.

Chapter 9

1. McNamara, K.J. and Scott, J.K. 1983. A new species of *Banksia* (Proteaceae) from the Eocene Merlinleigh Sandstone of the Kennedy Range, Western Australia. *Alcheringa* **7**: 185–193.
2. Jones, W.G., Hill, K.D. and Allen, J.M. 1995. *Wollemia nobilis*, a new living Australian genus and species in the Araucariaceae. *Telopea* **6**: 173–176.
3. Crane, P.R., Friis, E.M. and Pedersen, K.R. 1995. The origin and early diversification of angiosperms. *Nature* **374**: 27–33.
4. Sun, G., Guo, S.-X., Zheng, S.-L., Piao, T.-Y. and Sun, X.-K. 1993. First

discovery of the earliest angiospermous megafossils in the world. *Science in China*, Series B, **36**: 249–256.
5. Hill, C.R. 1996. A plant with flower-like organs from the Wealden of the Weald (Lower Cretaceous), southern England. *Cretaceous Research* **17**: 27–38.
6. Taylor, D.W. and Hickey, L.J. 1990. An Aptian plant with attached leaves and flowers: implications for angiosperm origin. *Science* **247**: 702–704.
7. Pellmyr, O. 1992. Evolution of insect pollination and angiosperm diversification. *Trends in Ecology and Evolution* **7**: 46–49.
8. Labandeira, C.C., Dilcher, D.L., Davis, D.R. and Wagner, D.L. 1994. Ninety-seven million years of angiosperm–insect association: paleobiological insights into the meaning of coevolution. *Proceedings of the National Academy of Science, USA* **91**: 12278–12282.
9. Rozefelds, A.C. and Sobbe, I. 1987. Problematic insect leaf mines from the Upper Triassic Ipswich Coal Measures of southeastern Queensland, Australia. *Alcheringa* **11**: 51–57.
10. Labandeira, C.C. and Phillips, T.L. 1996. A carboniferous insect gall: insight into early ecologic history of the Holometabola. *Proceedings of the National Academy of Science, USA* **93**: 8470–8474.
11. Shear, W.A. 1991. The early development of terrestrial ecosystems. *Nature* **351**: 283–289.
12. Edwards, D., Selden, P.A., Richardson, J.B. and Axe, L. 1995. Coprolites as evidence for plant–animal interaction in Siluro-Devonian terrestrial ecosystems. *Nature* **377**: 329–331.
13. Retallack, G.J. 1997. Early forest soils and their role in Devonian global change. *Science* **276**: 583–585.
14. McElwain, J. and Chaloner, W.G. 1995. Stomatal density and index of fossil plants track atmospheric carbon dioxide levels in the Palaeozoic. *Annals of Botany* **76**: 389–395.

Chapter 10

1. Taylor, M.A. 1997. Before the dinosaur: the historical significance of the fossil marine reptiles. *In* J.M. Callaway and E.L. Nicholls (eds), *Ancient Marine Reptiles*, pp. xix–xlvi, Academic Press, San Diego and London.
2. Storrs, G.W. 1997. Morphological and taxonomic clarification of the genus *Plesiosaurus*. *In* J.M. Callaway and E.L. Nicholls (eds), *Ancient Marine Reptiles*, pp. 145–190, Academic Press, San Diego and London.
3. Bakker, R. 1993. Jurassic sea monsters. *Discover*, **September 1993**: 78–85.
4. Long, J.A. 1998. *Dinosaurs of Australia and New Zealand, and other Creatures of the Mesozoic Era*. University of New South Wales Press, Sydney: Harvard University Press, Cambridge, Massachusetts.

5. Cruikshank, A.I. 1997. A lower Cretaceous pliosaurid from South Africa. *Annals of the South African Museum* **105**: 207–226.
6. Kauffman, E.G. and Kesling, R.V. 1960. An Upper Cretaceous ammonite bitten by a mosasaur. *Contributions of the Museum of Paleontology, University of Michigan* **15**: 193–248.
7. Russell, D.A. 1967. Systematics and morphology of American mosasaurs. *Peabody Museum of Natural History, Yale University Bulletin* **23**: 1–241.
8. Caldwell, M.W. and Lee, M.S.Y. 1997. A snake with legs from the marine Cretaceous of the Middle East. *Nature* **386**: 705–709.
9. Lee, M.S.Y. 1997. The phylogeny of varanoid lizards and the affinities of snakes. *Philosophical Transactions of the Royal Society of London* **B352**: 53–91.

Chapter 11

1. Wellnhoffer, P. 1991. *The Illustrated Encyclopedia of Pterosaurs*. Crescent Books, New York.
2. Kellner, A.W.A. and Langston Jr, W. 1996. Cranial remains of *Quetzalcoatlus* (Pterosauria, Azdarchidae) from Late Cretaceous sediments of Big Bend National Park, Texas. *Journal of Vertebrate Paleontology* **16**: 222–231.
3. Lehman, T.M. and Langston Jr, W. 1996. Habitat and behaviour of *Quetzalcoatlus*: palaeoenvironmental reconstruction of the Javelina Formation (Upper Cretaceous), Big Bend National Park, Texas. *Journal of Vertebrate Paleontology* **16**(3) abstracts: 48A.
4. Martill, D.M. 1997. From hypothesis to fact in a flight of fantasy: the responsibility of the popular scientific media. *Geology Today* **March/April 1997**: 71–73.
5. Sharov, A.G. 1971. New flying reptiles from the Mesozoic of Kazakhstan and Kirgizia. *Akademia Nauk, Paleontological Institute, Trudy* **130**: 104–113 (in Russian).
6. Unwin, D.M. and Bakhurina, N.N. 1994. *Sordes pilosus* and the nature of the pterosaur flight apparatus. *Nature* **371**: 62–64.
7. Padian, K. 1983. Osteology and functional morphology of *Dimorphodon macronyx* (Buckland) (Pterosauria: Rhamphorhynchoidea) based on new material in the Yale Peabody Museum. *Postilla* **189**: 1–44.
8. Bennett, S.C. 1997. Terrestrial locomotion of pterosaurs: a reconstruction based on *Pteraichnus* trackways. *Journal of Vertebrate Paleontology* **17**: 104–113.
9. Padian, K. 1984. The origin of pterosaurs. *In* W. Reif and F. Wastphal (eds), *Third Symposium on Mesozoic Terrestrial Ecosystems*, pp. 163–168, Attempto Verlag, Tübingen.

10. Bramwell, C.D. and Whitfield, G.R. 1974. Biomechanics of *Pteranodon*. *Philosophical Transactions of the Royal Society of London, Biological Sciences* **B267**: 503–592.
11. Bennett, S.C. and Long, J.A. 1991. A large pterodactyloid pterosaur from the Late Cretaceous (Late Maastrichtian) of Western Australia. *Records of the Western Australian Museum* **15**: 435–444.
12. Buffetaut, E., Laurent, Y., LeLoeuff, J. and Bilottes, M. 1997. A terminal Cretaceous giant pterosaur from the French Pyrenees. *Geological Magazine* **134**: 553–556.
13. Bennett, S.C. 1996. Year classes of pterosaurs from the Solnhofen Limestone of Germany: taxonomic and systematic implications. *Journal of Vertebrate Paleontology* **16**: 432–444.

Chapter 12

1. Hammer, W.R. and Hickerson, W.J. 1994. A crested theropod dinosaur from Antarctica. *Science* **264**: 828–830.
2. Long, J.A. 1998. *Dinosaurs of Australia and New Zealand, and other Creatures of the Mesozoic Era*. University of New South Wales Press, Sydney; Harvard University Press, Cambridge, Massachusetts.
3. Rich, T.H. and Rich, P.V. 1989. Polar dinosaurs and biotas of the Early Cretaceous of southeastern Australia. *National Geographic Research* **5**: 15–53.
4. Warren, A.A., Kool, L., Cleeland, M., Rich, T.H. and Vickers-Rich, P. 1991. An Early Cretaceous labyrinthodont. *Alcheringa* **15**: 327–332.
5. Rich, P.V., Rich, T.H., Wagstaff, B.E., McEwen Mason, J., Douthitt, C.B., Gregory, R.T. and Felton, E.A. 1988. Evidence for low temperatures and biologic diversity in Cretaceous high latitudes of Australia. *Science* **242**: 1403–1406.
6. Brush, A., Martin, L.D., Ostrom, J. and Wellnhoffer, P. 1997. Bird or dinosaur? – Statement of a team of specialists. *Episodes* **20**: 47.
7. Chen, P., Dong, Z. and Zhen, S. 1998. An exceptionally well-preserved theropod dinosaur from the Yixian Formation of China. *Nature* **391**: 147–152.
8. McKelvey, B. 1994. Cold war over warm ice. *Australian Natural History* **24**: 48–53.
9. Jell, P.J. and Roberts, J. (eds) 1983. Plants and invertebrates from the Lower Cretaceous Koonwarra Fossil Bed, South Gippsland, Victoria. *Memoirs of the Australasian Association of Palaeontologists* **3**: 1–205.
10. Molnar, R.E., Flannery, T.F. and Rich, T.H. 1981. An allosaurid theropod dinosaur from the Early Cretaceous of Victoria, Australia. *Alcheringa* **5**: 141–146.

11. Welles, S.P. 1983. *Allosaurus* (Saurischia, Theropoda) not yet in Australia. *Journal of Paleontology* **57**: 196.
12. Molnar, R.E., Flannery, T.F. and Rich, T.H. 1985. Aussie *Allosaurus* after all. *Journal of Paleontology* **59**: 1511–1513.
13. Rich, T.H. and Vickers-Rich, P. 1994. Neoceratopsians and ornithomimosaurs: dinosaurs of Gondwana origin? *National Geographic Research and Exploration* **10**: 129–131.
14. Currie, P.J., Vickers-Rich, P. and Rich, T.H. 1996. Possible oviraptosaur (Theropoda, Dinosauria) specimens from the Early Cretaceous Otway Group of Dinosaur Cove, Australia. *Alcheringa* **20**: 73–79.
15. Rich, T.H. 1996. Significance of polar dinosaurs in Gondwana. *Memoirs of the Queensland Museum* **39**: 711–718.

Chapter 13

1. Chiappe, L.M. 1995. The first 85 million years of avian evolution. *Nature* **378**: 349–355.
2. Barthel, K.W., Swinburne, N.H.M. and Conway Morris, S. 1990. *Solnhofen: A Study in Mesozoic Palaeontology*. Cambridge University Press, Cambridge.
3. Thulborn, R.A. and Hamley, T.L. 1985. A new palaeoecological role for *Archaeopteryx*. *In* M.K. Hecht, J.H. Ostrom, G. Viohl and P. Wellnhofer (eds), pp. 81–89, *The Beginnings of Birds*, Proceedings of the International Archaeopteryx Conference, Eichstätt, 1984.
4. Welman, J. 1995. *Euparkeria* and the origin of birds. *South African Journal of Science* **91**: 533–537.
5. Novas, F.E. and Puerta, P.F. 1997. New evidence concerning avian origins from the Late Cretaceous of Patagonia. *Nature* **387**: 390–392.
6. Thulborn, R.A. 1985. Birds as neotenous dinosaurs. *Records of the New Zealand Geological Survey* **9**: 90–92.
7. Norell, M.A., Clark, J.M., Demberelyin, D., Rhinchen, B., Chiappe, L.M., Davidson, A.R., McKenna, M.C., Altangerel, P. and Novacek, M.J. 1994. A theropod dinosaur embryo and the affinities of the Flaming Cliffs dinosaur eggs. *Science* **266**: 779–782.
8. Martin, L.D. 1991. Mesozoic birds and the origin of birds. *In* H.-P. Schultze and L. Trueb (eds), pp. 485–540, *Origins of the Higher Groups of Tetrapods*, Cornell University, Ithaca.
9. Hou, L.-H., Zhou, Z., Martin, L.D. and Feduccia, A. 1995. A beaked bird from the Jurassic of China. *Nature* **377**: 616–618.
10. Padian, K. 1996. Early bird in slow motion. *Nature* **382**: 400–401.
11. Sanz, J.L., Chiappe, L.M., Rérez-Moreno, B.P., Moratalla, J.J., Hernández-Carrasquilla, F., Buscalioni, A.D., Ortega, F., Poyato-Ariza,

F.J., Rasskin-Gutman, D. and Martínez-Delclòs, X. 1997. A nestling bird from the Lower Cretaceous of Spain: implications for avian skull and neck evolution. *Science* **276**: 1543–1546.
12. Sanz, J.L., Chiappe, L.M., Rérez-Moreno, B.P., Buscalioni, A.D., Moratalla, J.J., Ortega, F. and Poyato-Ariza, F.J. 1996. An early Cretaceous bird from Spain and its implications for the evolution of avian flight. *Nature* **382**: 442–445.
13. Hou, L., Martin, L.D., Zhou, Z. and Feduccia, A. 1996. Early adaptive radiation of birds: evidence from fossils from northeastern China. *Science* **274**: 1164–1167.
14. Chiappe, L.M. 1996. Late Cretaceous birds of southern South America: anatomy and systematics of Enantiornithines and *Patagopteryx deferrariisi*. *Münchner Geowissenschaftiche Abhandlungen* A **30**: 203–244.
15. Altangerel, P., Norell, M.A., Chiappe, L.M. and Clark, J.M. 1993. Flightless bird from the Cretaceous of Mongolia. *Nature* **362**: 623–626.
16. Chinsamy, A., Chiappe, L.M. and Dodson, P. 1995. Mesozoic avian bone microstructure: physiological implications. *Paleobiology* **21**: 561–574.
17. Feduccia, A. 1996. *The Origin and Evolution of Birds*. Yale University Press, Yale.
18. Cooper, A. and Penny, D. 1997. Mass survival of birds across the Cretaceous–Tertiary boundary: molecular evidence. *Science* **275**: 1109–1113.
19. Boles, W.E. 1997. Fossil songbirds (Passeriformes) from the Early Eocene of Australia. *Emu* **97**: 43–50.

Chapter 14

1. Coria, R.A. and Salgado, L. 1995. A new giant carnivorous dinosaur from the Cretaceous of Patagonia. *Nature* **377**: 224–226.
2. Long, J.A. and McNamara, K.J. 1995. Heterochrony in dinosaur evolution. *In* K.J. McNamara (ed.), *Evolutionary Change and Heterochrony*, pp. 151–168, Wiley, Chichester and New York.
3. Bakker, R.T., Williams, M. and Currie, P. 1988. *Nanotyrannus*, a new genus of pygmy tyrannosaur, from the latest Cretaceous of Montana. *Hunteria* **1**(5): 1–30.
4. Long, J.A. and McNamara, K.J. 1997. Heterochrony: the key to dinosaur evolution. *In* D. Wolberg (ed.), *Dinofest International*, Proceedings of a symposium sponsored by Arizona State University, pp. 113–123, The Academy of Natural Sciences, Philadelphia.
5. Varricchio, D.J. 1993. Bone microstructure of the Upper Cretaceous theropod *Troodon formosus*. *Journal of Vertebrate Paleontology* **13**: 99–104.

6. Chinsamy, A. 1992. Ontogenetic growth of the dinosaurs *Massospondylus carinatus* and *Syntarsus rhodesiensis. Journal of Vertebrate Paleontology* **12**: 23A.
7. Geist, N.G. and Jones, T.D. 1996. Juvenile skeletal structure and the reproductive habits of dinosaurs. *Science* **272**: 712–714.
8. Barretto, C., Albrecht, R.M., Bjorling, D.E., Horner, J.R. and Wilsman, N.R. 1993. Evidence of growth plate and growth of long bones in juvenile dinosaurs. *Science* **262**: 2020–2023.
9. Carpenter, K.J. 1992. Tyrannosaurids (Dinosauria) of Asia and North America. *In* N. Mateer and Chen Pei-ji (eds), *Aspects of Nonmarine Cretaceous Geology*, pp. 250–268, China Ocean Press, Beijing.
10. Buffetaut, E., Suteethorn, V. and Tong, H. 1996. The earliest known tyrannosaur from the Lower Cretaceous of Thailand. *Nature* **381**: 689–691.
11. Ruben, J.A., Jones, T.D., Geist, N.R. and Hillenius, W.J. 1997. Lung structure and ventilation in theropod dinosaurs and early birds. *Science* **278**: 1267–1270.
12. Horner, J.R. and Currie, P.J. 1994. Embryonic and neonatal morphology and ontogeny of a new species of *Hypacrosaurus* (Ornithischia, Lambeosauridae) from Montana and Alberta. *In* K. Carpenter, K.F. Hirsch and J. Horner (eds), *Dinosaur Eggs and Babies*, pp. 312–336, Cambridge University Press, Cambridge.
13. Horner, J.R. 1984. The nesting behaviour of dinosaurs. *Scientific American* **250**: 130–137.
14. Horner, J.R. and Weishampel, D.B. 1988. A comparative embryological study of two ornithischian dinosaurs. *Nature* **332**: 256–257.
15. Galton, P.M. 1990. Stegosauria. *In* D. Weishampel, P. Dodson, and H. Osmolska (eds), *The Dinosauria*, pp. 435–455, University of California Press, Berkeley.
16. Martin, V. 1994. Baby sauropods from the Sao Khua Formation (Lower Cretaceous) in northeastern Thailand. *Gaia* **10**: 147–153.

Chapter 15

1. Archer, M., Hand, S.J. and Godthelp, H. 1991. *Riversleigh. The Story of Animals in Ancient Rainforests of Inland Australia.* Reed Books International, Sydney.
2. Archer, M., Flannery, T.F., Ritchie, A. and Molnar, R.E. 1985. First Mesozoic mammal from Australia – an early Cretaceous monotreme. *Nature* **318**: 363–366.
3. Flannery, T.F., Archer, M., Rich, T.H. and Jones, R. 1995. A new family of monotremes from the Cretaceous of Australia. *Nature* **377**: 418–420.

4. Pascual, R., Archer, M., Jaureguizar, E.O., Prado, J.L., Godthelp, H. and Hand, S.J. 1992. First discovery of monotremes in South America. *Nature* **356**: 704–706.
5. Godthelp, H., Archer, M., Cifelli, R., Hand, S.J. and Gilkeson, C.F. 1992. Earliest known Australian Tertiary mammal fauna. *Nature* **356**: 514–516.
6. Rich, T.H., Vickers-Rich, P., Constantine, A., Flannery, T.F., Kool, L. and van Klaveren, N. 1997. A tribosphenic mammal from the Mesozoic of Australia. *Science* **278**: 1438–1442.
7. Woodburne, M.O. and Case, J.A. 1996. Dispersal, vicariance, and the Late Cretaceous to Early Tertiary land mammal biogeography from South America to Australia. *Journal of Mammalian Evolution* **3**: 121–161.
8. Hedges, S.B., Parker, P.H., Sibley, C.G. and Kumar, S. 1996. Continental breakup and the ordinal diversification of birds and mammals. *Nature* **381**: 226–229.
9. Sereno, P.C. and McKenna, M.C. 1995. Cretaceous multituberculate skeleton and the early evolution of the mammalian shoulder girdle. *Nature* **377**: 144–147.
10. Thewissen, J.G.M. 1994. Phylogenetic aspects of cetacean origins: a morphological perspective. *Journal of Mammalian Evolution* **2**: 157–184.
11. Thewissen, J.G.M., Hussain, S.T. and Arif, M. 1994. Fossil evidence for the origin of aquatic locomotion in archaeocete whales. *Science* **263**: 210–212.
12. Gingerich, P.D., Raza, S.M., Arif, M., Anwar, M. and Zhou, X. 1994. New whale from the Eocene of Pakistan and the origin of cetacean swimming. *Nature* **368**: 844–847.
13. Thewissen, J.G.M. and Fish, F.E. 1997. Locomotor evolution in the earliest cetaceans: functional model, modern analogues, and paleontological evidence. *Paleobiology* **23**: 482–490.

Chapter 16

1. McNamara, K.J. 1994. The significance of gastropod predation to patterns of evolution and extinction in Australian Tertiary echinoids. *In* B. David, A. Guille, J.P. Firal and M. Roux (eds), *Echinoderms through Time (Echinoderms Dijon)*, pp. 785–793, Balkema, Rotterdam.
2. Kitchell, J.A., Boggs, C.H., Kitchell, J.F. and Rice, J.A. 1981. Prey selection by naticid gastropods: experimental tests and application to the fossil record. *Paleobiology* **7**: 533–552.
3. Chesher, R.H. 1969. Contributions to the biology of *Meoma ventricosa* (Echinoidea: Spatangoida). *Bulletin of Marine Science* **19**: 72–110.

4. McNamara, K.J. 1990. Echinoids. *In* K.J. McNamara (ed.), *Evolutionary Trends*, pp. 205–231, Belhaven, London.
5. Jablonski, D. and Bottjer, D.J. 1988. Onshore–offshore evolutionary patterns in post-Palaeozoic echinoderms: a preliminary analysis. *In* R.D. Burke, P.V. Mladenov, P. Lambert and R.L. Parsley (eds), *Echinoderm Biology*, pp. 81–90, Balkema, Rotterdam.
6. Stanley, S. 1979. *Macroevolution, Pattern and Process.* Freeman, San Francisco.
7. Donovan, D.K. and Gale, A.S. 1990. Predatory asteroids and the decline of the articulate brachiopods. *Lethaia* **23**: 77–86.

Chapter 17

1. McNamara, K.J. 1997. *Shapes of Time*. Johns Hopkins University Press, Baltimore.
2. Shea, B.T. 1989. Heterochrony in human evolution: the case for human neoteny. *Yearbook of Physical Anthropology* **32**: 69–101.
3. White, T.D., Suwa, G. and Asfaw, B. 1994. *Australopithecus ramidus*, a new species of early hominid from Aramis, Ethiopia. *Nature* **371**: 306–312.
4. Jablonski, N.G. and Chaplin, G. 1993. The origin of habitual terrestrial bipedalism in the ancestor of the Hominidae. *Journal of Human Evolution* **24**: 259–280.
5. Smith, B.H. 1991. Dental development and the evolution of life history in Hominidae. *American Journal of Physical Anthropology* **86**: 157–174.
6. Aiello, L.C. and Wheeler, P. 1995. The expensive-tissue hypothesis. *Current Anthropology* **36**: 199–221.
7. Bermúdez de Castro, J.M., Arsuaga, J.L., Carbonell, E., Rosas, A., Martínez, I. and Mosquera, M. 1997. A hominid from the Lower Pleistocene of Atapuerca, Spain: possible ancestor to neandertals and modern humans. *Science* **276**: 1392–1395.
8. Gibson, K.R. 1991. Myelination and behavioral development: a comparative perspective on questions of neoteny, altriciality and intelligence. *In* K.R. Gibson and A.C. Petersen (eds), *Brain Maturation and Cognitive Development*, pp. 29–64, De Gruyter, New York.
9. Ruff, C.B., Trinkaus, E. and Holliday, T.W. 1997. Body mass and encephalization in Pleistocene *Homo*. *Nature* **387**: 173–176.
10. Parker, S.T. and Gibson, K.R. 1979. A developmental model for the evolution of language and intelligence in early hominids. *Behavioral and Brain Science* **2**: 367–408.

Chapter 18

1. MacFadden, B.J. 1986. Fossil horses from '*Eohippus*' (*Hyracotherium*) to *Equus*: scaling, Cope's Law, and the evolution of body size. *Paleobiology* **12**: 355–369.
2. Reilly, S.M. 1994. The ecological morphology of metamorphosis: heterochrony and the evolution of feeding mechanisms in salamanders. *In* P.C. Wainwright and S.M. Reilly (eds), *Ecological Morphology: Integrative Approaches in Organismal Biology*, pp. 319–338, University of Chicago Press, Chicago.
3. Reilly, S.M. 1987. Ontogeny of the hyobranchial apparatus in the salamanders *Ambystoma talpoideum* (Ambystomatidae) and *Notophthalmus viridescens* (Salamandridae): the ecological morphology of two neotenic strategies. *Journal of Morphology* **191**: 205–214.
4. Osborn, H.F. 1929. The titanotheres of ancient Wyoming, Dakota, and Nebraska. *US Geological Survey Monograph* 55.
5. McKinney, M.L. and Schoch, R.M. 1985. Titanothere allometry, heterochrony, and biomechanics: revising an evolutionary classic. *Evolution* **39**: 1352–1363.
6. Bales, G.S. 1996. Heterochrony in brontothere horn evolution: allometric interpretations and the effect of life history scaling. *Paleobiology* **22**: 481–495.
7. Oakley, K.P. 1981. Emergence of higher thought 3.0–0.2 Ma B.P. *Philosophical Transactions of the Royal Society of London* **B292**: 205–211.
8. Smith, W.G. 1894. *Man the Primeval Savage*. Edward Stanford, London.
9. Oakley, K. 1965. Folklore of fossils. Part II. *Antiquity* **39**: 117–125.
10. Oakley, K. 1965. Folklore of fossils. Part I. *Antiquity* **39**: 9–16.
11. McNamara, K.J. and Philip, G.M. 1980. Australian Tertiary schizasterid echinoids. *Alcheringa* **4**: 47–65.
12. McNamara, K.J. 1990. The role of heterochrony in evolutionary trends. *In* K.J. McNamara (ed.), *Evolutionary Trends*, pp. 59–74, Belhaven, London.
13. Smith, A.B. 1984. *Echinoid Palaeobiology*. George Allen & Unwin, London.

Chapter 19

1. Agosti, D., Grimaldi, D. and Carpenter, J.M. 1997. Oldest known ant fossil discovered. *Nature* **391**: 447.
2. Sloan, R.E. 1997. Plate tectonics and the radiations/extinctions of dinosaurs, the Pele Hypothesis. In D.L. Wolberg, E. Stump and G.

Rosenberg (eds) *Dinofest International: Proceedings of a Symposium held at Arizona State University*, pp. 533–539, Academy of Natural Sciences, Philadelphia.

3. Hengst, R.A., Rigby, J.K., Landis, G.P. and Sloan, R.E. 1995. Biological consequences of Mesozoic atmospheres. In N. MacLeod and G. Keller (eds) *The Cretaceous-Tertiary Mass Extinction: Biotic and Environmental Events*, pp. 327–347, Norton, New York.
4. Ruben, J.A., Jones, T.D., Geist, N.R. and Hillenius, W.J. 1997. Lung structurc and ventilation in theropod dinosaurs and early birds. *Nature* **278**: 1267–1270.

Index

Note: page numbers in italics denotes illustration